Future Flight
The Next Generation of Aircraft Technology
2nd Edition

To the hope that the machines of Future Flight will be used for the benefit of mankind and not its destruction.

Light from the past shines on the present and lights the pathway into the future.

John Busick, 1987

Future Flight
The Next Generation of Aircraft Technology
2nd Edition

Bill Siuru, Ph.D.
John D. Busick, M.A.O.M.

TAB **AERO**
Division of McGraw-Hill, Inc.
Blue Ridge Summit, PA 17294-0850

pbk   10 11 12   QPF/QPF   0 5 4
hc   1 2 3 4 5 6 7 8 9 10   QPF/QPF   9 9 8 7 6 5 4 3

Library of Congress Cataloging-in-Publication Data

Siuru, William D.
    Future flight: the next generation of aircraft technology / Bill
Siuru, John D. Busick. — 2nd ed.
       p.   cm.
    Includes index.
    ISBN 0-8306-4377-X (h)   ISBN 0-8306-4376-1 (p)
    1. Airplanes—Technological innovations.   I. Busick, John D.
II. Title.
TL671.S565   1993
629.133'34—dc20   93-32089
                                                              CIP

Acquisitions Editor: Jeff Worsinger
Editorial team: Joanne Slike, Executive Editor
                        Lori Flaherty, Managing Editor
                        Pam Reichert, Book Editor
Production Team: Katherine G. Brown, Director
                        Rhonda E. Baker, Typesetting
                        Jan Fisher, Typesetting
                        Susan E. Hansford, Typesetting
                        Ollie Harmon, Typesetting
                        Lisa M. Mellott, Typesetting
                        Brenda M. Plasterer, Typesetting
                        Wendy Small, Pagination
                        Cindi Bell, Proofreading
Book Design: Jaclyn J. Boone
Cover design: Carol Stickles, Allentown, Pa.   AV1
Photograph courtesy of NASA   4367

Contents

Acknowledgments

ACA Industries, Inc.
 Dr. Julian Wolkovitch
 Paul Fjeld

Aerospatiale
 Jerome Rondeau

Airbus Industrie

American Blimp Corporation
 James R. Thiele

Aurora Flight Sciences Corporation
 Alice Ann Toole

Avco Lycoming

Basler Turbo Conversions, Inc.
 Warren L. Basler

Beech Aircraft Corp.
 John Gedraitis

Bell Aerospace Textron

Bell Helicopters Textron
 Terry Arnold
 Bob Leder

Boeing Commercial Airplane Co.
 Maureen M. Herward
 Stephanie Bauer

Boeing Helicopters
 Jack Satterfield

Boeing Military Airplanes Co.
 John Kvasnosky

British Aerospace
 Trevor Mason

Cadillac Motor Car Division
 Charles Harrington

Canadair

Cirrus Design Corporation
 Tim Tamcsin
 Brad Zeman

Department of the Air Force
 Bill Holder
 Helen Kavanaugh
 Jo Anne Rumple

Deutsche Forschungsanstalt für Luft
und Raumfahrt

Flight Link Flight Simulator Controls
  Mary Henkenius

General Dynamics/Forth Worth

General Electric
  Ken Kilner

Grumman Aerospace Corp.

Gulfstream Aerospace Corp.

Hughes Helicopters

IAI International Inc.
  Debby Shoham

Learjet, Inc.

Litton Systems Canada Ltd.
  C. W. Pittman

Lockheed Aeronautical Systems
Company
  Ramonda Siniard

Lockheed Georgia

Luftschiffbau Zepplin GmbH
  M. Mugler/K. Hagenlocher

Marquardt & Roche Co.
  Michael Figoff

Martin Marietta

McDonnell Douglas Helicopters
  Hal Klopper

NASA/Ames Research Center
  Robert H. Stroub

NASA/Langley Research Center

NASA/Lewis Research Center
  Linda Lewis

Northrop Corporation
  C. John Amrhein

Northwest Airlines
  Brenda K. Parsons

Pratt & Whitney (United Technologies)
  Carole Jenkins

Rockwell International/MBB

Rockwell International/Collins
Avionics & Communications
  Jim Thebeau

Rotec

Rutan Aircraft Factory

Saab Military Aircraft
  Asa Holm

Seawind

U.S. Army Aviation and Troop
Command
  Robert E. Hunt

Westinghouse Electronic Systems Group
  Ann Grizzel
  Bryan Wiggins

Special thanks to Brian Siuru for proofreading and correcting the final manuscript.

Preface

SINCE THE BEGINNING of history man has yearned to reach for the heavens. Long before flight was thought possible, the wings of eagles, appearing on the art of ancient Egypt, the staffs of Roman legions, and the Great Seal of the United States, represented man's thirst for flight. American Indians revered the eagle because it represented the lofty ideal of their spirits leaving the bonds of earth and soaring to the heavens.

When flight became possible, the eagle symbology was still there. Charles Lindbergh was known as the "Lone Eagle." And who will ever forget the transmission from the moon, "the Eagle has landed?" Today, one of the U.S. Air Force's finest air superiority fighters is the F-15 Eagle. The aircraft of the future will combine the spirit of the eagle with the ingenuity of man.

The past three decades have seen military and commercial aircraft routinely fly at more than twice the speed of sound. Travel by air has progressed from a means of transportation for the privileged, and for special occasions, to the most popular means of traveling distances greater than a few hundred miles. And travel into space, while not yet routine, is a familiar event.

As an engineer in advancing technology and as a military pilot using these advances in technology, we have not only watched the rapid progress of aviation, we have had an active part in it. This intimate experience with what has happened in the past, along with the knowledge of what is going on now in the aerospace world, allows us to project what will be happening in the future.

Introduction

THE DEVELOPMENT of most aviation concepts usually follows a rather orderly process from the gleam in the eyes of the engineer and scientist to the final aircraft in the hands of the military, commercial, or private pilot. Whether the aircraft is a fast fighter or a luxurious executive transport, the process starts with a need. The need can be the quest to conquer a new frontier in flight or simply do an existing transportation job better and more economically.

Technology is the most important ingredient in satisfying these needs. Some technology is already waiting in the laboratory, waiting for a "home" in a new aircraft. In other cases, technological barriers must be surmounted by additional work of scientists and engineers. The technology comes together in a complete aircraft that meets the original need. This whole process can take years—even decades—in the case of sophisticated military aircraft and commercial airliners. Thus, by looking at the technology now being developed and adding an insight into the needs of the future, a picture of the aircraft of the twenty-first century emerges.

In this book we will take this sequence—needs, technology, concepts—to predict the aircraft of the future. Because history is an important part in understanding the future, the book will start with a look at lessons that can be learned from the past.

While the theory and mathematics of flight can quickly become very complicated, the basic principles are usually quite simple to understand. We have attempted to explain the technology and concepts in terms that anyone with only an interest in aviation can comprehend.

Much has happened in aviation technology since the first edition of *Future Flight* was published in 1987. Some of the technology has matured. It is now already included or being incorporated in the next generation of aircraft. Other technology has been dropped because it did not provide the expected results, was not feasible or practical, or was just too expensive. This second edition has been updated to cover advances in aerodynamics, propulsion systems, avionics, materials, artificial intelligence, and manufacturing techniques. For instance, major advances have occurred in technologies like neural networks, supercomputers, vectored thrust, composites, fiber optics, and much more.

The future was greatly impacted by the end of the Cold War, the collapse of the Soviet Empire, and the resulting downsizing of the U.S. military forces. This second edition discusses how changes in the world have changed the future of military, commercial, and general aviation.

These world-shaking changes will have the greatest effect on military aviation technology. Already military systems are taking a different course with new systems for new missions and new roles for old systems. This book presents up-to-date information on the B-2 Stealth Bomber, F-22 fighter, RAH-66 Comanche helicopter, and other military aircraft.

Civilian aviation is also progressing rapidly with both new concepts and revivals of old ones now made possible with major advances in electronics, propulsion, and materials. For instance, new material is included on the latest Boeing, McDonnell Douglas, and Airbus airliners.

1

Lessons from the past

DECEMBER 17, 2003 will mark an important date in the history of man: the centennial of powered flight. Aviation has come a long way from the eventful day in 1903 when the Wright brothers proved man could fly. Before this 100th anniversary is celebrated, aviation will have progressed even further. This book will provide a look at these advances as well as give a glimpse of the aircraft that could be flying the skies in 2003. You'll also discover some of the aviation technology engineers and scientists are developing today for use even further into the future.

PREDICTING THE FUTURE

Most advances in air travel have evolved with small steps forward rather than giant leaps. Indeed, the truly revolutionary inventions like the jet engine, atomic energy, the microchip, the laser, and the transistor are rare. Therefore, in many instances, predictions for the future start with projecting what evolutionary improvements will develop from what is flying in the skies today.

Few concepts are totally new. Even the Wright brothers' early planes had a canard, a pusher propeller, and tricycle landing gear—features considered modern today. You cannot thumb through many early post-World War II issues of *Popular Mechanics* or *Popular Science* without coming across ideas for plastic airplanes. It took the development of sophisticated composites, however, to make the plastic airplane possible. Even the supersonic combustion ramjet (scramjet), the possible means of propelling the National Aerospace Plane (X-30), was fairly well developed during the 1960s before it fell into disfavor, mainly because of production difficulties and cost. The B-2 Stealth bomber uses a flying-wing concept that is reminiscent of the Northrop XB-35 and YB-49 flying wings of the 1940s (Fig. 1-1). In the future, you will see a revival of concepts from the past now made possible by advances in aerodynamics, materials, propulsion, electronics, and manufacturing technology.

Many of the advances in aerospace technology started with one man's invention. Take, for instance, Frank Whittle's and Pabst von Ohain's jet engines, Michel Wibault's vectored-thrust engine, Robert Jones' scissor wing, Julian Wolkovitch's joined wing, or T. H. Maiman's laser. Today, scientists at universities and research

Fig. 1-1. The Northrop flying wing of the 1940s was, perhaps, ahead of its time. However, it would appear again a few decades later as the central design of stealth-type aircraft like the B-2 bomber. This is a Northrop YB-49.

laboratories are working on theories that will be the basis of aerospace systems of the future. Looking at their discoveries provides another insight into the future.

In the future, as in the past, the major burden of developing completely new aircraft will be borne by governments. In the United States, this means the military services and the National Aeronautics and Space Administration (NASA) will be the leaders in development. Other major developments will be undertaken by consortiums of aerospace companies. Their efforts will take on even greater importance as development costs are measured in billions of dollars.

No private aerospace company, regardless of size, has the resources to take on the ambitious development projects that can mean bankruptcy if unsuccessful. Some projects are so expensive even governments won't be able to go it alone, resulting in partnerships between nations. This is not to say that all developments will be at the national or international level. There is still a place for individual breakthroughs with promising financial rewards for companies, especially in the electronics, materials, and computer industries.

In the past, research aircraft like the X planes were used to try out new ideas before they were incorporated into production aircraft, usually military aircraft (Table 1-1). The military's and NASA's X planes of the 1940s and 1950s provided much information about flying at high altitudes and at transonic and supersonic speeds (Fig. 1-2). Other experimental aircraft, like the M2-F1/F2/F3, HL-10, and X-24A/B, were specifically built to investigate lifting body concepts (i.e., aircraft that obtain lift from their bodies rather than from conventional wings), and thus provided valuable data for the Space Shuttle. The world of Vertical/Short Take-off and Landing (V/STOL) aircraft is filled with many experimental aircraft, like

Table 1-1. Significant experimental research aircraft: The X planes.

Aircraft	Manufacturer	Years flown	Major goal	Max. speed (mph)	Max. altitude (ft.)
D-558 I	Douglas	1947 - 1953	Investigate flight characteristics at transonic speeds	650.8	40,000
D-558 II	Douglas	1948 - 1956	Investigate swept-wing aircraft at high supersonic speeds	1291.0	83,235
X-3	Douglas	1952 - 1955	Investigate flight at sustained supersonic speeds	> Mach 1	41,318
X-1A/B/ D/E	Bell	1953 - 1958	Investigate flight at higher speeds and altitudes	1612.0	90,440
X-4	Northrop	1948 - 1953	Investigate semi-tailless aircraft at high subsonic speeds	620.0	40,000
X-5	Bell	1950 - 1954	Investigate aircraft capable of sweeping wings in flight	716.0	49,919
XF-92A	Convair	1951 - 1953	Investigate delta-wing aircraft at transonic speeds	Mach 1	42,464
X-15	North American	1959 - 1968	Investigate flight at very high speeds and altitudes	4520.0	354,200
X-29A	Grumman	1985 -	Investigate flight of forward-sweep aircraft	—	—
XB-70	North American	1965 - 1969	Investigate flight at high speeds	Mach 3	70,000

Fig. 1-2. The most famous of the X planes has to be the X-1A in which Chuck Yeager broke the sound barrier.

Fig. 1-3. One of the many less-than-successful attempts at a V/STOL aircraft, the Bell X-22A.

the XPY-1, X-13, XV-4A/B, and XV-3, that were less than unqualified successes (Fig. 1-3).

Today, building experimental aircraft strictly for research purposes has become cost prohibitive. The first new X plane to come along in more than a decade was the Grumman X-29A designed to investigate forward-sweep wing aerodynamics (Fig. 1-4). The Rockwell International/Messerschmitt-Bolkow-Blohm X-31A was aimed at flight testing concepts for future highly-maneuverable military fighters (Fig. 1-5).

Fig. 1-4. The Grumman X-29A is one of the latest aircrafts in the X plane "family."

4 Lessons from the past

Fig. 1-5. The Rockwell International/Messerschmitt-Bolkow-Blohm X-31A is another modern X plane designed to demonstrate Enhanced Fighter Maneuverability technology.

It is no wonder then that some of today's experimental aircraft are actually highly modified versions of aircraft already in production. The U.S. Air Force's very successful Advanced Fighter Technology Integration (AFTI) program used a highly modified General Dynamics F-16 Fighting Falcon. NASA has used a variety of modified commercial and military transports for experimental platforms. For example, a Boeing 737 became a Terminal-Configured Vehicle (TCV) for studying ways to make landings safer, especially in severe weather. In the Aircraft Energy Efficiency (ACEE) program, modified air transports like the KC-135 and DC-10 have flight tested new aerodynamic, propulsion, and flight control ideas aimed at saving fuel. The USAF is now flying the Variable Stability In-Flight Simulator Test Aircraft (VISTA). This modified F-16D can be programmed to mimic the flying qualities of new aircraft not yet built.

Scale models have been used to flight test new concepts. Models range from small radio-controlled models much like the ones built by hobbyists to models that are manned. The Beech Starship I general aviation aircraft made its first flight as an 85-percent scale model and the Fairchild T-46A, at one time a concept for a new Air Force primary trainer, first flew as a 62-percent scale model.

Remotely-piloted research vehicles (RPRVs) offer another way to flight test new concepts without the huge expense of man-rating research aircraft. One very successful example was the joint Air Force/NASA Highly Maneuverable Aircraft Technology (HiMat) craft that tested ideas for use on high-performance fighter aircraft (Fig. 1-6). Other RPRVs were used during the development of the McDonnell Douglas F-15 and General Dynamics F-16 and to evaluate new vertical takeoff and landing V/STOL configurations.

Fig. 1-6. Remotely piloted research vehicles like this HiMat unmanned aircraft allow flight testing without the need for expensive man-rated aircraft.

Computers are making great contributions in the development of new aircraft. Complete aircraft can be mathematically modeled on a computer, providing data that before could only be obtained from costly and time-consuming flight and wind tunnel testing. Furthermore, changes to designs can be made simply by hitting a few keystrokes or moving a light pen.

As in the past, much of the technology will filter down from the high-speed fighters to military and commercial transports. Some of this technology can even be incorporated into general aviation craft, just as the jet engine (first used in fighters) became practical on transports like the de Havilland Comet and Boeing 707 and eventually on business jets like the Gates Learjet and Cessna Citation. Today, you can see a diffusion of technology in such things as composite materials that were first used on military fighter aircraft and are now being used on the latest business aircraft. Or you can see the F-16's sidestick controller now found in the cockpits of new commercial airliners, such as the European Airbus.

TRENDS FOR THE FUTURE

It is safe to say that there are no new physical frontiers in air travel. With the Space Shuttle, man has flown almost the entire flight envelope in terms of speed and altitude. Helicopters and V/STOL aircraft can fly forwards, backwards, sideways, and even stay suspended in midair. Aircraft have been flown in all sizes, from the tiny man-powered ultralights to the giant Lockheed C-5 and Boeing 747. Incidentally, the last decade does not have a monopoly on aircraft size. Aircraft like the Sikorsky Bolshoi bomber (1913), Dornier Do X (1929), Messerschmitt Gigant (1941), and Saunders-Roe Princess (1962) were all huge aircraft.

So where do the challenges lie for the rest of this century and the next? For military fighters and bombers, the key requirement will be the ability to survive in wartime and accomplish their missions in the face of an increasingly sophisticated enemy threat. This translates into building aircraft with greater maneuverability, better defensive systems, and stealth technology. And future military designs have to be economical enough to build and operate in sufficient quantity.

While flying faster and with more passengers is of interest to the airlines, it will only happen if it means greater profits and can be done safely. Airlines have learned hard lessons about operating aircraft that are at the leading edge of technology or have too much capacity. The jumbo jets and the supersonic transport (SST) are prime examples of this. Future airliner development will be aimed at greater fuel efficiency, lower operating costs, and most important, safety. While you might see some exotic airliners like the hypersonic transport by the end of the 21st century, most air travel in the early part of the century will still be on airliners that look much like those used today. However, attached to (and within) the rather conventional-looking fuselages will be an abundance of new materials, engines, electronics, and other high-technology items.

Advances in electronics will probably be the leading factor influencing future military and commercial transport aircraft. Lasers, computer-generated graphics, voice control, high-speed computer chips, and artificial intelligence will all play an important role. Space satellites already provide important weather and navigation information to pilots in real time and are becoming a key part of air traffic control in skies that will become increasingly crowded. With all this information available to both military and commercial pilots, considerable thought will have to be given to equipment and techniques that reduce the pilot's workload.

General aviation encompasses everything that is not military or commercial aviation, including business aircraft, crop dusters, light and sport aircraft, even ultralights and homebuilts. Today, business aircraft like the Beech Starship I are at the forefront of aircraft design and technology. New agricultural aircraft concepts are being designed, and nowhere in aviation will you find a greater array of innovative designs than in the aircraft being built in basements and garages across America.

For the private pilot, the upwardly spiraling cost of punching holes in the sky is a great problem. For aircraft flown mainly for pleasure, low initial and operating costs will be the driving force. Some of the technology developed for military and commercial aircraft will filter down to private aircraft, either as it becomes cheap enough or is mandated by safety regulations. Composite materials are already the mainstay of the homebuilder, and there are many Cessnas, Pipers, and even Luscombes built decades ago that have sophisticated electronics retrofitted to their instrument panels.

Our prediction of the future begins by looking at the key technologies being developed in university, private, and government laboratories around the free world. Then we will look at various military, commercial, and general aviation aircraft that will make use of this technology in the next half century.

Our look into the future is limited to the technologies and concepts possible and feasible based on current scientific knowledge. No wild fantasies and no ideas that violate the laws of gravity, thermodynamics, or nature are presented. Some concepts, however, will be pushing these laws to their limits. Any prediction of the future is limited by the fact that we cannot forecast the effect of unknown breakthroughs in science. Who could have predicted, even 20 years ago, the impact discoveries like the computer chip and lasers would have on the world today?

AVIATION TECHNOLOGY TODAY

Before we begin our journey into the future, let's look at the current state of aerospace technology as a benchmark for measuring future progress. We'll briefly examine the latest representatives of military, commercial, and general aviation aircraft flying today and see what new technologies they are already using. A word of caution here. While these aircraft might represent the latest things in the sky, their technologies are often dated. Because of the long lead time required to develop new aircraft, the technology they incorporate might be 5, 10, or even more years old already, even though it was state-of-the-art when the design was frozen and the fabrication was started.

Northrop's B-2

Northrop's Advanced Technology Bomber is designed for long-range heavy bombing with a variety of nuclear and conventional weapons, including SCRAM missiles, "smart" bombs, gravity bombs, and maritime weapons (Fig. 1-7). While the B-2 looks nothing like any aircraft flying in the sky today, it does look more than a little like Northrop's previous XB-35 and YB-49 flying wing bombers. However, advances in technology now make the flying wing a practical aircraft.

The reason for the B-2's unique shape is, of course, because its flying wing provides excellent stealth characteristics. Indeed, the B-2 is the first bomber built from the start using stealth principles. Its shape plus other areas of stealth technology give the B-2 the capability of penetrating the most sophisticated enemy air defenses at all altitudes up to 50,000 feet.

The four General Electric jet engines buried in the upper wings can push the 376,000 pound B-2 to high subsonic speeds. The B-2 has a range of 6,000 miles without refueling, which can be extended to 10,000 nautical miles with one refueling. The prototype made its maiden flight on July 17, 1989. Building the B-2 required the development of nearly 900 new materials and manufacturing processes. Many of these developments, some considered breakthroughs, are being transferred to other segments of American industry.

Fig. 1-7. Northrop brought back the flying wing for the B-2 bomber. Unlike the 1940s versions, new technology made the flying wing work, and work well.

Lockheed's F-22

The U.S. Air Force's Advanced Tactical Fighter (ATF) is destined to replace the venerable F-15 in air superiority missions. When the F-22 is in service it will be the Air Force's first new air superiority aircraft in two decades, and will probably have to serve as the Air Force's first-line fighter into the year 2010 or even 2020 (Fig. 1-8)!

Air superiority means being the winner in a dogfight requiring designers to strike the right balance between acceleration, advanced avionics, low-observability, and maneuverability. The F-22 combines stealth technology (so it is virtually invisible to enemy detection sensors) and surface-to-surface missiles with the superior maneuverability needed to be the "top gun" in air-to-air combat. The F-22's twin Pratt & Whitney F119-PW-100 turbofan engines provide the bursts of acceleration needed for dogfighting without the use of fuel guzzling afterburners.

Fig. 1-8. The Lockheed F-22 Advanced Tactical Fighter is destined to be the USAF's air superiority fighter well into the twenty-first century.

Afterburners can increase fuel usage up to 500 percent when they are lit as well as provide a huge heat signature for infrared heat-seeking missiles to target. The F-22 will be able to "supercruise" at about Mach 1.5 without afterburning. The F-22 engines also include up-and-down thrust-vectoring of the engine's hot exhaust, allowing the F-22 to fly at very high angles of attack to further enhance maneuverability.

By the time the F-22 goes into production in the mid-1990s, 40 percent of the aircraft will be constructed of advanced composite materials. The F-22's pilot flies behind full-color, multifunction, liquid crystal displays in the cockpit.

Boeing's 777

The 777 is a medium-size commercial airliner whose capacity falls somewhere between the Boeing 767-300 and 747-400 and is the world's largest twinjet. With a fuselage diameter of 20⅓ feet, it is wider than any other jetliner except the Boeing 747 jumbo jet (Fig. 1-9). The 777 can be configured in any number of seating arrangements from six to ten seats abreast to provide capacity from 305 to 328 passengers in triple-class setups and as many as 440 passengers in all-economy seating configurations. The 777 has a range of up to nearly 5,000 statute miles, with growth capability of up to 7,600 miles. The cruising speed is Mach 0.83.

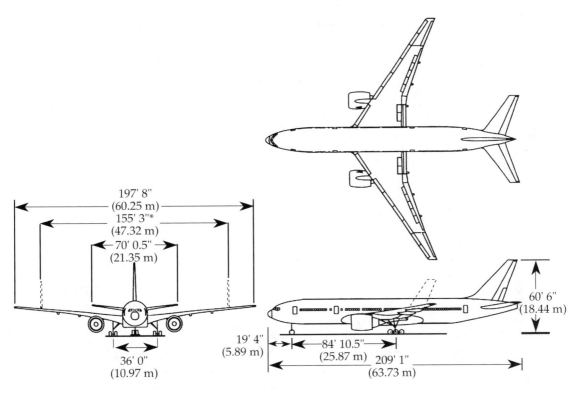

197' 8"
(60.25 m)
155' 3"*
(47.32 m)
70' 0.5"
(21.35 m)

36' 0"
(10.97 m)

19' 4"
(5.89 m)

84' 10.5"
(25.87 m)

209' 1"
(63.73 m)

60' 6"
(18.44 m)

*Optional wingtip fold shown.

Fig. 1-9. The Boeing 777 is the world's largest twinjet airliner. Boeing Commercial Airplanes.

The Boeing 777 uses the latest in aerospace technology including composites, plus the latest avionic technology including large flat-panel color displays in the cockpit, and a fly-by-wire flight control system.

Gulfstream's IV and V

The Gulfstream IV and V high-technology business jets represent the upper level of capability available in the general aviation segment. The Gulfstream IV-SP (Special Performance) uses twin Rolls-Royce Tay Mk611-8 turbofan engines. The Gulfstream IV-SP can cruise at Mach 0.8 with a Mach 0.88 maximum speed. While capable of carrying up to 19 passengers, this business aircraft has a maximum range of over 3,000 miles carrying three crew members, eight passengers, and a fuel load that allows a maximum weight landing.

Because business jet users now want non-stop, global-range capability between all major business and government centers, the Gulfstream V will be in operation by 1996 (Fig. 1-10). Its range is 7,200 miles, allowing businessmen to fly from New York to Buenos Aires, Honolulu to London, or Denver to Hong Kong non-stop, remaining airborne for up to 14 hours. The Gulfstream V flies at speeds

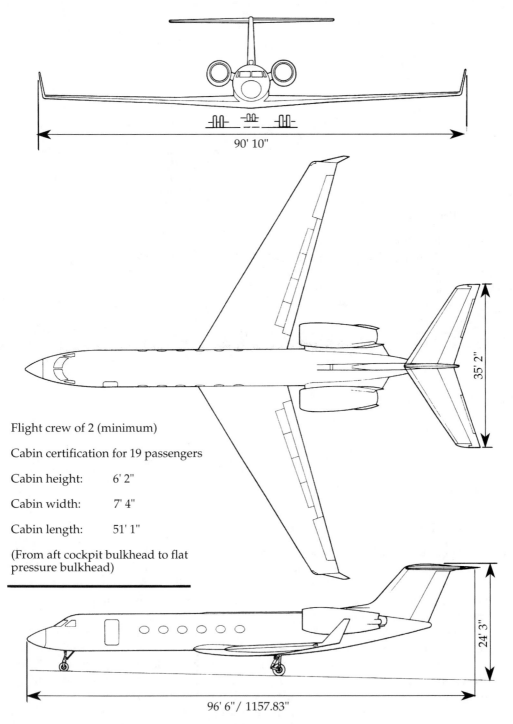

90' 10"

35' 2"

Flight crew of 2 (minimum)

Cabin certification for 19 passengers

Cabin height: 6' 2"

Cabin width: 7' 4"

Cabin length: 51' 1"

(From aft cockpit bulkhead to flat
pressure bulkhead)

24' 3"

96' 6" / 1157.83"

Fig. 1-10. When it comes to general aviation aircraft, business jets like this Gulfstream V are the
top-of-the-line. Gulfstream Aerospace Corporation.

of up to Mach 0.9 and at altitudes of up to 51,000 feet. Its power comes from twin BMW/Rolls-Royce BR 710 engines.

Because passengers aboard business aircraft usually work while in flight, Gulfstreams are equipped with an airborne office that includes fax machines, an IBM PS/2 computer workstation, and satellite communications capabilities. Other ammenities can include a super galley, crew rest area, exercise area, and lavatories that can be equipped with showers (Fig. 1-11). The Gulfstreams have collision avoidance and turbulence prediction equipment as well as the latest navigation systems.

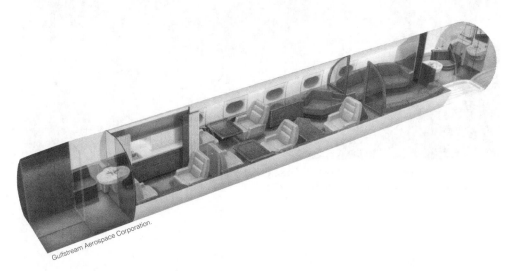

Gulfstream Aerospace Corporation.

Fig. 1-11. "First Class" accommodations are a must aboard business aircraft like the Gulfstream V.

British Aerospace/Aerospatiale Concorde

The Concorde is the only supersonic transport flying today (Fig. 1-12). A total of 14 Concordes have been built and are flying for British Airways and Air France. Despite the fact that they were built prior to 1979 and are based on technology now more than a quarter of a century old, Concordes are still advanced aircraft. They have to be because they fly as fast as 1500 MPH. The Concorde's tiny windows get as hot as oven doors and the fuselage actually stretches about nine inches. Not even military jets fly at supersonic speeds for as long as the Concorde does.

The world's other two supersonic transports did not progress as well as the Concorde. The Soviets withdrew their TU-144 from service after only a brief period of operation, and the domestic SST never got off the drawing board.

Fig. 1-12. The Concorde has proven that a supersonic transport is quite feasible even though it is currently used on very limited routes.

2

Future aviation needs

THERE IS A SAYING in the aerospace industry: "Requirements push and technology pulls." What this means is that the requirements of new missions, or even the need to do current jobs better, drive engineers and scientists to work on the leading edge of technology (Fig. 2-1) to find solutions to problems posed by ever more demanding requirements through the invention and development of new ideas and technology. New concepts are constantly being invented in university, government, and commercial laboratories. It takes forward-thinking planners to envision how these new technologies can be used for improving military and civilian aerospace capabilities and to actually develop new aircraft around these breakthroughs in technology. This is how "technology pulls."

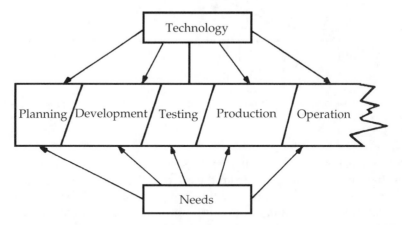

Fig. 2-1. An aircraft's life cycle, from initial conception to retirement, can last decades. The development of a new airplane, especially a military craft, can last 10, or even 20 years. And as experience with the B-52 and C-130 has shown, a basic design might be in operation for half a century or more.

This chapter examines the missions and jobs military and civilian aircraft must do in the future and presents the requirements for advancing technology. Subsequent chapters examine the technology being pursued to produce aircraft that can meet these requirements, as well as new technology that could result in aircraft as yet unimagined.

MILITARY MISSIONS

Military missions place the greatest demands on technology. This is easily seen by the vast sums of money required to develop new weapon systems like the B-2 bomber and the Air Force's F-22 Advanced Tactical Fighter (ATF) (Fig. 2-2). Even a relatively unsophisticated aircraft like the C-17 military airlift transport represents a rather large investment.

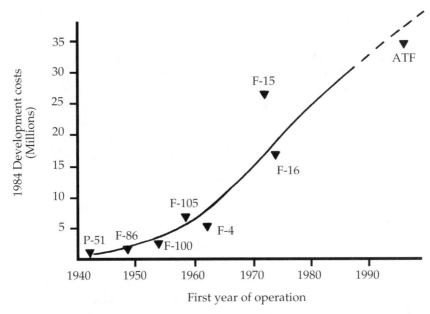

Fig. 2-2. Aircraft development costs have grown drastically through the years. This shows the flyaway costs for typical Air Force fighters.

The U.S. Air Force, Navy, Army, and Marine Corps are assigned a variety of missions that must be conducted from the air, or at least require support from airborne craft. The Air Force is naturally the "big gun" in the aerospace world, with missions ranging from strategic bombing and air-to-air combat to transporting vast numbers of troops anywhere in the world on a moment's notice. Added to this are the important jobs of reconnaissance and surveillance.

The Navy's air requirements are focused on the defense of its fleet, protection of the submarine force, protection of naval power ashore, and supporting U.S. naval forces around the world. Both the Air Force and Navy are needed to give close air support to troops fighting on the ground.

The Army's aircraft provide direct support of ground troops. Thus, the Army relies heavily on helicopters for attacking targets, transporting troops in the field, and getting supplies to the troops. Except for strategic bombing and reconnaissance, Marine Corps aviation does what the Air Force, Navy, and Army do, but on a more limited scale.

The past few years have brought about significant political changes. U.S. military aerospace development was driven for decades by the threat of the Soviet Union's military and technological capabilities. Now the Soviet Union has collapsed. But as witnessed by events like the Persian Gulf War, the revolution in the former Yugoslavia, and the humanitarian mission in Somalia, the world is still a dangerous place. Furthermore, the United States, as the only remaining superpower, will probably be called upon frequently to help resolve armed conflicts throughout the globe.

In the 1990s, the U.S. military will undergo some major changes. Factors like a reduced Soviet threat yet an increased threat from more capable and aggressive Third World nations mean a complete revision of the military's missions and roles. Further complicating the problem is a reordering of national priorities from defense to domestic issues, leading to significantly smaller defense budgets and a downsized military force.

While the military planners have to conceive new missions as well as revised strategies and tactics to meet new world conditions, they are constrained by the fact that for the most part they have to do the job with existing hardware. This not only means adapting weapon and support systems already in the inventory, but also upcoming systems like the B-2 bomber and F-22 fighter. The systems in which development started years ago will probably fall by the wayside if they cannot be adapted for new roles or are too expensive for much smaller budget constraints.

Air superiority fighters

"The gaining of air superiority is the first requirement for the success of any major land operation." This quote is as appropriate today as when it appeared in a 1943 U.S. army field manual. In today's uncertain military planning environment, developing new weapons systems to counter very vague future air superiority threats is difficult and froth with the potential for drastic mistakes . . . mistakes that we would have to live with for perhaps a half a century. For example, the development of the Advanced Tactical Fighter was started in the early 1980s, but it will not be in the field in significant quantities until after the year 2000. Based on past experience with today's frontline fighters, it will probably have to be used for 35 years or more. All this adds up to an aircraft that was designed for an original threat that occurred 50 years ago. Can you imagine if the dogfights of the Korean Conflict were fought with aircraft designed for the needs of World War I, such as Spads and Fokkers!

As an air superiority fighter, the F-22 is a heavy multipurpose fighter incorporating state-of-the-art electronics and providing a long combat range, the ability to carry a large armament load, and the ability to fly and fight under all weather conditions. It has to do all this without sacrificing the ability to survive both in air-to-air combat and against surface-to-air threats. This all adds up to a need for speed, stealth, and maneuverability (Fig. 2-3).

Fig. 2-3. YF-22 Advanced Tactical Fighter shown being refueled by a KC-135 tanker. The F-22 will undoubtedly remain the USAF's primary air superiority fighter well into the twenty-first century.

When the initial requirements for the Advanced Tactical Fighter were outlined in the early 1980s, the future threat was clear—Soviet Union fighters would be ready by the mid-1990s and would fly well into the twenty-first century. In the 1980s, the Soviets were modernizing their fighter fleet at a much faster pace than the United States. The F-22 was designed to eventually replace the F-15, a very sophisticated and capable aircraft in its own right.

With the collapse of the world's only other superpower, the threat changed drastically and is now not very clear. Indeed, some ask if a very advanced and expensive aircraft like the F-22 is really needed. However, the F-22 should still be built in relatively large quantities for a couple of very valid reasons.

First, (perhaps remote, but still worth considering) is that the Soviet Union may rise again. The Soviets will undoubtedly continue to develop and field advanced fighters like the Su-27 and MiG-29, albeit in much smaller quantities.

Even more important, the F-22 would be ready to take on well-equipped Third World adversaries equipped with high-technology weapons. Ever more lethal weapons can be found in the air forces of Third World nations. Beside the Soviet-built aircraft already mentioned, other nations like France, Germany, Great Britain, and even Japan will be offering advanced fighters on the world's arms market. Thus the F-22 might have to counter advanced fighters like the French Rafale or the European Fighter that are now under development. Moreover, if the F-15 and F-16 experience is repeated, the F-22 will be sold to other nations. Our pilots could conceivably engage other F-22's in the hands of a country that has become an adversary. The Third World battlefield could be as demanding as anything that might have occurred in Central Europe. The F-22 is a hedge against the uncertainties of the future. It is a good insurance policy.

Bombers

Long-range, strategic bombers like the B-1B, B-2, and yes, even the venerable B-52, will continue to be important in the future. First, strategic bombers will still be a key in the U.S.'s Triad deterrence policy, along with the USAF's land-based intercontinental ballistic missiles (ICBMs) and the Navy's submarine-launched ballistic missiles (SLBMs). Triad is the concept of using ICBMs, SLBMs, and strategic bombers as a three-pronged deterrence to Soviet nuclear aggression. Each of the three Triad elements contributes its own unique deterrence threat as a result of its individual characteristics and means of deployment. The strategic bomber's contribution to Triad is its unique capability to be recalled after launch. It can be used as a show of power and recalled before actually inflicting damage. It is flexible in the sense that its designated targets can be changed while the aircraft is enroute, an impossibility with both the ICBM and SLBM, whose targets are programmed prior to launch. Because of its "man-in-the-loop" feature, it is considered the most stabilizing of the three Triad elements. Finally, the bomber is the only one of the three that can be used for other missions, such as non-nuclear, conventional bombing in limited war situations.

Perhaps a bigger question is whether or not Triad is needed now at the end of the Cold War and its accompanying threat of United States-Soviet nuclear confrontation. The answer is yes, but at a reduced level. The world is still an uncertain place and the military direction the former Soviet Union will take has yet to be fully determined. Even without any new weapon systems, the Soviets will continue to possess the capability to destroy the United States within 30 minutes in the foreseeable future. Moreover, if the Soviets continue to modernize and improve the effectiveness of their strategic nuclear arsenal at the current pace, the result would be a totally modernized Soviet Triad by the turn of the century. While arms control agreements will result in reductions of deployed nuclear weapons, current limits still allow the Soviets to act on their military objectives.

When the B-2 was still called the Advanced Strategic Penetrating Aircraft, its mission statement called for the capability to conduct missions across the spectrum of conflict, including general nuclear war, conventional conflict, and peace-time/crisis situations. Of course, during the Cold War the greater emphasis was on the bomber's nuclear capability as part of the Triad. Reduced United States-Soviet tensions and the current volatile Middle East situation have resulted in emphasizing the B-2's conventional role. Conventional bombing capabilities of strategic bombers have already been proven by the B-52. During the Vietnam war the B-52 flew only 5 percent of the total sorties, yet delivered 50 percent of the total tonnage. In Operation Desert Storm, B-52s dropped 30 percent of the total tonnage.

Finally, as the United States downsizes its military forces and withdraws from overseas bases, long-range bombing capability assumes even greater importance. Aircraft with range and payload capability are needed to provide a display of force in the first days of a crisis that can occur anywhere in the world.

The need for stealth characteristics for bombers was dramatically demonstrated during the Persian Gulf War. The stealthy F-117A was able to operate with impunity even in heavily defended areas, and without the need for large escort aircraft or expensive techniques to suppress enemy defenses. The B-2 can fly on the same unobserved mission with 5 to 6 times more payload and to ranges 5 to 6 times greater than the F-117.

Close air support

All four military services are called upon to provide close air support to ground troops. The Air Force, Navy, and Marine Corps use fixed-wing aircraft for this role, while the Army and Marine Corps use helicopters for ground-support work.

Close-air-support fighter requirements include low observables and the ability to operate from damaged runways, and for the Navy and Marine Corps, from ships with flight decks much smaller than a typical aircraft carrier. Aircraft of all types will probably have to operate from bomb-cratered runways in any future war. This means aircraft should require a minimum of real estate, something the Navy already knows well from its carrier-based operations.

The ideal close-support fighter is one that can operate from airfields or ships located near the forward edge of the battle area (FEBA). The idea is to have aircraft that can quickly fly back to support bases for rearming and refueling and return to the thick of battle within minutes. This implies the need for vertical/short takeoff and landing capability. Because ground battles continue in the worst of weather and at night, the close-support fighter must have all-weather and night fighting capability.

The close-support aircraft must also be survivable. First, it must be able to avoid detection. That is where low observables come into play. Then, if it is detected, it must be able to survive hits from enemy fire, mainly in the form of small-arms fire. The aircraft must be able to fly home with battle damage and then be easily repaired for a quick return to combat. Finally, to be able to fight effectively, the ground-support pilot will require assistance in handling his tremendous workload.

Airlift transports

Airlift transports are very large aircraft needed to carry large contingents of troops and their fighting equipment anywhere in the world. Recent events like the Persian Gulf War and the Somalia effort again reinforce the need for long-range airlift capability.

Not only must these aircraft be able to carry large tonnage, they must be able to load, carry, and unload outsized cargo, such as the tanks, helicopters, and howitzers that are a part of the modern arsenal of weapons. High subsonic cruising speeds are desirable in airlift transports in order to get the maximum number of troops to a trouble spot in the minimum amount of time.

The demands for airlift during a major crisis usually far exceeds the capability of the military airlift fleet. Therefore, civilian air transports are called to active duty to airlift troops and cargo. While these military requirements do not drive the design of commercial airliners, they do have to be considered.

Assault transports

Whereas the airlift transports get the troops and equipment to the general region of combat, the assault transport must take these men and their gear to the exact point on the battlefield where they are needed. This can be done by airborne assault using paratroops or by landing and unloading right on the battlefield. In either case, the assault transport must be able to operate from the most primitive runways—airstrips that will usually be damaged. It must also be essentially self-sufficient, requiring no ground support equipment or ground-based navigation aids. Naturally, survivability in combat is a requirement. Helicopters, vertical takeoff and landing aircraft, and short takeoff and landing aircraft are all natural approaches to satisfying these requirements.

Reconnaissance and surveillance

Space satellites perform reconnaissance and surveillance today, but there will always be a need for "spy" aircraft that fly in the atmosphere, or at least at its very fringes. Cloud obscuration, poor weather, and the laws of orbital mechanics sometimes mean that needed information cannot always be obtained from orbiting spacecraft. High- and fast-flying reconnaissance aircraft must be ready to do the job. These aircraft will have to fly over enemy territory without being detected, and stealth technology is a must.

Tactical reconnaissance and surveillance for actual combat operations is very time-crucial because of the changeability of the modern battlefield. The military cannot totally rely on satellite information, but must send in reconnaissance aircraft to obtain close-up and up-to-the-minute information on enemy positions, buildups, and electronic order of battle. Tactical reconnaissance aircraft must have low observables and be able to survive enemy action. It must also be able to operate from primitive and battle-damaged airfields.

Because many reconnaissance and surveillance missions are dangerous, unmanned drones and remotely-piloted vehicles (RPVs) are a consideration. RPVs are less expensive than manned aircraft and are expendable.

Aircraft that are part airplane and part space satellite are also generating interest for both military and civilian applications. These ultrahigh-altitude, unmanned craft can fly for weeks, or perhaps even months at a time, with very low fuel consumption. Some even have the ability to fly on solar power.

Trainers

Aircraft needed to train new military pilots range from primary and basic trainers to advanced trainers used for combat practice. Until 1960, the U.S. Air Force used a "dual track" to train its pilots. After initial training, pilots moved on to either single-engine or multiengine trainers depending on whether they were going to eventually fly fighters, bombers, or transports. Since 1960, a single track Undergraduate Pilot Training (UPT) program has been used where pilots get primary training in the Cessna T-37, and then go on to the Northrop T-38 for advanced training even if they were eventually going to fly multiengine aircraft. Specialized training in line with their first operational aircraft was not given until the very end of the training program.

Today, the Air Force is returning to the "dual track" training program to better prepare new pilots in the aircraft they will fly on their first assignment after UPT. After completing a common initial training, student pilots proceed along either the bomber-fighter or tanker-transport tracks. The program needed new aircraft because the Air Force did not have a multiengine, side-by-side trainer. The USAF is purchasing the T-1A Jayhawk, essentially a modified Beechcraft 400A twin-engined business jet, for this purpose.

Replacements for both the T-37 and T-38 are included in the Air Force's long-range plans. The Air Force will first replace the T-37 with a new primary trainer called the *Joint Primary Aircraft Trainer System* (JPATS), which will also be used by the Navy to replace the Beech T-34.

For initial instruction, an aircraft must be safe and economical to operate, rather forgiving of student mistakes, and able to take a lot of punishment, yet be able to perform high-performance aerobatics. Trainers must also use little fuel and require minimum maintenance, even though flight hours are accumulated rapidly. Further in the future, probably around the year 2005, the USAF will have a bomber-fighter training system for training bomber-fighter pilots.

In the future, more training will be done in craft that never leave the ground, that is, in simulators. Simulators are far less expensive and can fabricate situations that could never be attempted in a real airplane.

Rotorcraft

A rotorcraft is a hybrid aircraft that combines the features of both helicopters and fixed-wing aircraft. The rotorcraft offers unique military capabilities. Vietnam is often referred to as the "Helicopter War" because of the tremendous contributions made by helicopters like the Bell-built Huey transports and gunships. The future of combat helicopters is very promising in light of the need to operate near the forward edge of the battle area (FEBA), often from completely destroyed runways. Added to this is the helicopter's unique ability to fly among trees and rocks to avoid detection. The helicopter not only can land troops on the battlefield, but can also return to extract them when the mission is over or when a retreat is ordered.

The helicopter has several shortcomings. One of these is that its top speed is limited to about 200 knots. Thus, the need exists for rotorcraft that also use other means (besides the rotors) for high-speed flight. Such devices might also improve yet another deficit of the helicopter, its range.

Because a helicopter has to work so close to the enemy, low observables and survivability against enemy action are acute problems. While great progress has been made in reducing the infrared, radar, visible, and the difficult acoustic signature of the helicopter, there is still much work to be done to achieve an "invisible" and "silent" helicopter. Other goals for future helicopter designs include increased payload capacity, reduced maintenance requirements, less vibration, and reduced pilot workload.

An interesting new mission is appearing on the horizon for the design of helicopters—surviving a dogfight. Dogfights between missile-armed and gun-armed helicopters are inevitable. Not only will this require developing new tactics, but also advancing technology, especially improving maneuverability and survivability.

Feasibility vs. affordability

While all sorts of new concepts and applications of technology are feasible, they are not all economically possible. First, there is the cost of development. Many ideas are much too expensive. Then there is the matter of life-cycle costs. Not only the original development cost must be considered, but the cost of buying operational aircraft and operating them on a day-to-day basis for decades. These operational costs can often far outshadow original development and procurement investments. Two of the most expensive items here are fuel costs and maintenance expenses. Future technological advances are needed in the less glamorous, but vitally important propulsion and ease-of-maintenance technologies.

COMMERCIAL AIR PASSENGER SERVICE

Military aircraft designs are driven by military mission requirements, and while the cost to buy and operate an aircraft is important, it is not the paramount consideration. For commercial aircraft, profitability is a major consideration, as is safety.

The universal parameter used by airlines in evaluating profitability is the direct operating cost per seat mile. The greatest cost of running an airline today is the cost of fuel. Fortunately, this cost can be reduced by advances in technology, if the cost savings are not outweighed by the investment required. To keep investments within bounds, many techniques to reduce fuel usage are rather simple ones, especially in comparison to technologies needed for military aircraft (e.g., stealth requirements). Fuel savings can come from more efficient engines, so there will be much emphasis on improving engine efficiency in the years to come. Fuel savings can also result from lower-drag designs as well as simple techniques like keeping and airplane's fuselage and wings clean. A few years ago a NASA study showed that a dirty Lockheed L-1011 could cost as much as $100,000 more per year to operate than a clean one—and that was when jet fuel was only 32 cents per gallon. Making aircraft lighter by using composite materials is another fuel-saving technique.

While not as dramatic as the effect of reduced fuel costs, lower personnel costs are also of interest to the airlines, particularly reducing the size of maintenance crews and flight crews. Advances in electronics, computers, and artificial intelligence could make a single-pilot airliner quite feasible, and indeed, even make a robotically flown aircraft possible. It is unlikely, however, that either of these will ever become reality; passengers will probably never feel comfortable unless at least two pilots man the cockpit. On a more positive note, future technology from military aircraft will find its way into the commercial airliner cockpit to reduce pilot workload so pilots can cope with flying in increasingly crowded skies.

Another way to improve profitability is to increase the number of seats on airliners, but this is more easily said than done. Drawing from past experience with jets like the Boeing 747, it is known that there are only a few routes in the world that generate profitable passenger load factors for jumbo airliners. The most promising routes are those covering the Pacific Rim where distances between destinations and the number of travelers are both great. To meet these needs, some aircraft builders are designing airliners with seating capacities of 600, 700, or more passengers. However, there are several challenges that might impede the use of such jumbo aircraft. For one thing, the huge development and purchase investments are difficult to amortize on those few profitable routes unless the aircraft are also used extensively for cargo-carrying duties. For the most part, airlines prefer the flexibility of smaller aircraft. Also, while seldom mentioned, the adverse publicity that would result by loss of life in the crash of a 500- to 700-passenger aircraft is another consideration.

Because of the huge cost of developing entirely new airliners, current airliners will be used well into the twenty-first century. Some of these are likely to be updated with new engines and minor aerodynamic changes to improve their efficiency. It is quite possible that some airframes will be used for a half-century with modifications and improvements (Fig. 2-4).

Boeing Military Airplanes Division.

Fig. 2-4. While future commercial airliners will constantly incorporate advances in technology, their external appearance will probably not change that much from today's airliners, like this Boeing 757.

Airliners rolling off assembly lines in the next few decades will be evolutionary models of airliners now in operation. Do not expect any radical changes in subsonic airliner design. Most people feel comfortable and safe flying in aircraft they are familiar with, especially those airlines that have built up safe and reliable records.

Many of the requirements for more economical, more efficient, and safer airliners will be satisfied by incorporating advances in computer and electronic technology. Such things as computerized engine and flight controls, as well as a myriad of high-technology communications, navigation, and safety devices will help meet the airlines' needs for improved operations within their financial constraints.

Electronics can often be installed on aircraft by adding a few black boxes and rerouting wiring. Some technology improvements require only the rewriting of computer software. This does not imply, however, that all electronic developments come cheaply. They still require a significant investment for research, development, testing, and certification.

If you take a casual look into the future, your conclusion might be that the aircraft of the twenty-first century will fly faster; indeed, some experts think they will fly at hypersonic speeds. This premise deserves a closer look, especially from a bottom-line economics standpoint.

There is no getting around the fact that faster airplanes are more costly to build and operate. When you fly at supersonic or hypersonic speeds, more sophisticated propulsion systems, materials, flight controls, and just about everything else are needed. Because of this, costs jump, sometimes by an order of magnitude. But where the costs really rise is in day-to-day operations. At supersonic speeds, engines guzzle tremendous amounts of fuel, and maintenance becomes much more complex because of the stresses and temperatures put on exotic materials.

All of this means that a fairly stiff premium must be paid by passengers who want to save a few hours in the air. As witnessed by British Airways' and Air France's experience with the Concorde, there are only a few routes in the world where there is sufficient passenger interest in paying a steep surcharge. This is a far cry from the original plans for Concorde, which showed it covering the globe. One reason is that supersonic flights over many countries, most significantly the United States, are prohibited because of sonic boom problems. Supersonic flights are usually constrained to transoceanic routes. The removal of these restrictions is highly unlikely, because of political and environmental considerations. Development of hypersonic airliners that fly at the fringes of the atmosphere would be a very expensive solution. Then there is the problem of flying in the stratosphere where nitrous oxide emissions from the engines can do serious damage to the earth's ozone layer. This concern for possible ozone depletion was one of the reasons for the cancellation of the United States's first supersonic transport. Today, the concerns for ozone layer depletion are even more acute.

Another consideration, often overlooked, is the frequency and convenience of flights. Saving two or three hours inflight is of little value to the busy executive if he or she must lose several hours because the airline's schedule doesn't fit the executive's schedule. Likewise, time gained enroute can be quickly lost if connecting flights mean long delays or if long-distance ground travel must be made to get to the selected airports where ultra-high-speed flights originate and terminate.

You shouldn't conclude, however, that supersonic and hypersonic airliners won't be developed for the twenty-first century. But they do present a much more difficult challenge if they are to be profitable.

In the next few decades, people will be traveling greater distances more routinely. The interweaving of business and commercial interests on a global basis will require long-distance business travel, especially in the Pacific region. Added to this is a projected increase in pleasure travel and an increase in the affluent retired population. High-speed aircraft are one way to reduce travel time. Another less technologically demanding approach is the development of longer-range aircraft that can reduce travel time by eliminating intermediate refueling stops.

Yet another way to reduce travel time is to reduce time spent in getting to and from airports, in boarding and disembarking, and in retrieving luggage. Then there is the time lost while in the holding pattern or waiting for takeoff clearance. Added to these are the security delays brought about by acts of terrorism that probably will, unfortunately, not diminish in the future. Solutions to these problems represent real payoffs in reducing travel time. But they are the most challenging problems airlines and airport operators have to face, indeed, more difficult than developing a faster airliner. And these problems will increase as airports and airways become even more congested.

Costs are not the only problems facing airlines today and in the future. High technology will be required to make air travel safer through improved traffic control, midair collision avoidance systems, and instrumentation for flights in poor weather conditions. Airlines are already aware of the pressure of environmental groups to alleviate noise and air pollution around airports.

Another important aspect of air travel is the short-hop commuter business. While air transportation already accounts for almost all long-distance travel (at least in the United States), commuter airlines must compete with ground transportation, especially the automobile. Today, distances of about 200 miles or less are most efficiently and cost effectively covered by car. To capture this segment of the market, air terminals must be built near city centers, requiring that advanced concepts, such as V/STOL aircraft, be pursued. Ironically, the poor fuel economy and high noise levels of V/STOL aircraft present another technology challenge.

Short range commuter aircraft will also face stiff competition from high speed and magnetic levitation, or maglev, trains. Currently, surface alternatives to commuter airliners are planned in several parts of the United States where there is a high volume of traffic, such as routes that connect New York and Washington, D.C., San Francisco, and Los Angeles, and the Dallas-Houston-Austin triangle in Texas.

In the past half century, the airline industry has seen an almost exponential increase in passengers, along with reduced travel times. At the same time, the cost of air travel has decreased continually. If inflation and the ever-decreasing buying power of the dollar are factored in, the decrease in the cost of air travel is really dramatic. In the future, both the speed and capacity of airliners will remain essentially constant. The challenge for the airlines will be to keep fares as low as possible while still earning a fair profit. What will dramatically change, however, is the total volume of passengers carried by air. This will present the challenge of safely handling a much larger amount of traffic, both in the air and on the ground.

COMMERCIAL AIR CARGO SERVICE

Air cargo traffic will grow also, especially if air shipping costs can be made competitive with trucks, trains, and ships (for overseas destinations). Unlike airlines, cargo carriers can reduce costs by using very large aircraft. Not only would cargo transporters be able to economically transport a large volume of freight on a single flight, but they could also carry oversized items like construction equipment and heavy manufacturing machinery.

To air cargo carriers, volume and weight carrying capacity are often more important than maximum speed. One interesting possibility is a large aircraft capable of carrying both cargo and passengers. Aircraft designed like the Air Force's C-5, in which people are carried on the top deck and cargo is carried below, could be very profitable for service even to the more remote areas of the world.

Operating ultra large transports, however, will require major changes in the ground handling of cargo. Efficient methods need to be devised to quickly load and unload large volumes of freight. Also, to save time, airports designed strictly for handling large airfreighters could be built away from congested passenger terminals. There is no point in shipping by air to save time if time is lost on the ground. This is an especially important factor if air transport is to be competitive with trucking in the United States.

Because airports to service ultra-large transports would require large tracts of land, and because a bulk of their flights would originate and terminate at coastal cities, seaplanes might be a very logical solution. You might see large flying boats, modern versions of the giants that opened up transoceanic flight in the 1930s and early 1940s.

In the past, freight carriers used cast-off airliners, but with the advent of highly profitable and efficient air-express overnight services, things have really changed. The requirements of these carriers could have a major impact on the commercial air transport industry in the future. As the term "express" implies, these carriers' unique service is based on speed. Overnight delivery anywhere in the United States is the rule, and now the service has been extended to Europe and Asia. Indeed, today's high-technology industries are highly dependent on the express carrier. Now it is not necessary to maintain large inventories of products or replacement parts in various locations. With air express they can be maintained at a single location, reducing inventory and facility costs. Because of its need for speed, the air-express carrier could even use new supersonic, or even a hypersonic, transport to provide overnight service just about anywhere in the world.

But speed is not the only ingredient that has made the air-express business what it is today. Absolute reliability is why most customers use the express carriers. If a passenger airliner is late or canceled, a few hundred people might be inconvenienced. A cancelled air-express flight could mean tens of thousands of critical items would not get there when they were needed. Production lines, even whole companies, could be shut down for the lack of key electronic or computer parts. To meet schedules, air-express carriers must fly in all types of weather including rain, snow, fog, and especially darkness. Overnight delivery means plenty of night flying. The requirements of overnight delivery are almost identical to those needed for the military's all-weather fighters.

Air-express carriers could profitably make use of the super-large airfreighter, especially for future transoceanic operations. Because passengers are not involved, landings and takeoffs could take place from isolated airfields or, most likely, from seaports. Such a large aircraft would probably be a flying boat. The aircraft would be designed for rapid loading and unloading, so that packages could be quickly transferred to smaller aircraft for continuation to their final destinations.

At the other end of the scale, the air-express carriers could benefit from V/STOL aircraft. These could provide rapid intercity transport, avoiding congested airport facilities. They have an advantage over helicopters in both speed and range. A V/STOL could fly hundreds of miles each night, landing and taking off from small terminals on the outskirts of towns.

Besides an exponential growth in air transport because of a global economy, air transport could also become an integral part of the manufacturing process. This has already been demonstrated with Cadillac Allante convertibles. Allante bodies are built by Pininfarina in Turin, Italy and then shipped via the "Allante Airbridge" to Detroit, Michigan where the drivetrain is installed and final assembly is completed (Fig. 2-5). The Airbridge is a fleet of modified Boeing 747s operated by Alitalia Airlines that can each carry 56 partially completed cars on the 3300 mile trip.

Fig. 2-5. Cadillac Allantes being loaded on a Boeing 747 "Airbridge Allante" that connects the assembly process between Turin, Italy and Detroit, Michigan.

GENERAL AVIATION

General aviation can be divided into two categories, flying done for commercial purposes (other than the major airlines and freight carriers) and flying done just for the fun of it. In the first category there are the crop dusters, business and executive transports, aircraft used in the offshore petroleum industry, plus a multitude of other uses that are best served by light aircraft. In the second category are the aircraft used for personal transport and pleasure flying.

While general aviation is often overlooked, it represents the biggest segment in many categories. The 220,000 general aviation aircrafts represent about 90 percent of all aircraft flying in the U.S. today. Air mileage of general aviation is roughly equal to commercial air carrier mileage and the number of people carried in general aviation is a third of that of U.S. air carriers.

The general aviation industry currently is depressed for a variety of reasons, so there are some real challenges for the next few decades. Again, economics is a driving factor. Buying and operating a light aircraft is out of the reach of many people who, in the past, could have afforded an airplane. Light aircraft prices have soared from about the price of a luxury car to the price of an expensive home, or even more. The cost of certification and spiraling liability insurance premiums, as well as the multitude of electronics required to fly in today's crowded skies, are the major contributors to these dramatic price increases. While advanced technology can help drive down costs, the significant changes needed to save the general aviation industry will probably not come from the drawing board and laboratory, but from changes in product liability laws and government regulations.

One way the aviation buff can keep flying is through the use of ultralights and homebuilts. Interestingly, these low-cost aircraft are developing some advanced technology of their own, especially in the areas of aerodynamics and materials. Some of this technology might flow upstream. The homebuilts offer a very inexpensive way to try out new ideas.

Because of airspace congestion and the concern for safety in the skies, aircraft electronics, or avionics, have become very sophisticated and expensive. Some of these devices are mandated by the Federal Aviation Administration; others are optional but highly desirable. The challenge for the electronics industry is to design and build "black boxes" that the average aviation enthusiast can afford.

One of the largest single users of helicopters is the offshore petroleum industry. As offshore oil rigs move farther offshore to service new oil fields, the time involved in commuting to and from them by helicopter will become excessive and too expensive, especially when crews are paid "portal-to-portal." The answer could be V/STOL aircraft.

3

Aerodynamics

AERODYNAMICS determine the overall appearance of an aircraft. However, unlike products such as automobiles or refrigerators, every part of an aircraft's external styling has a specific purpose. Nothing, except for perhaps the paint job, is there just because it looks good. On some aircraft, even the paint is chosen to perform a specific function, such as to reduce aircraft temperatures or to make it difficult for the enemy to detect.

In the early days of aviation, aircraft designers had only a rudimentary understanding of aerodynamics, and the designs were mostly the result of intuition and trial-and-error. Today, aerodynamics is a highly-developed science, and aircraft designs are the result of extensive testing in both the wind tunnel and on the computer. In the future, the computer will have an even more significant impact on the aerodynamic design of aircraft. With ultrahigh-speed "super" computers, the designer can try out new designs simply by making a few changes with a light pen on a computer screen. Expensive wind tunnel models and flight testing are still needed to evaluate final designs.

The aerodynamics of a particular aircraft are determined by the aircraft's purpose. Optimum performance, whether it be maximum speed or altitude, or the important consideration of fuel economy, is largely determined by an aircraft's aerodynamics. Of course, aerodynamic designs have to be balanced with the fact that an aircraft must have engines to propel it, it must be able to carry people or cargo, and it must land and take off. Through the years, inventors have come up with great proposals for aircraft designs that never went into production, either because they were too difficult or expensive to build, or because there were no materials available that could withstand the stress of flight over years of operation. As you will see, some of these "way-out" ideas of the past will be possible in the future because of the major advances being made in materials and manufacturing technologies.

WINGS

The thing that probably strikes you first when you look at any aircraft is the design of the wings, the components that provide most of the aircraft's lift. Future aircraft will use many wing designs that are already proven and in operation on current aircraft. But they will also use some advanced ideas that will greatly enhance performance.

Flying wings

Because wings are what give an airplane lift, and just about every other part of the airframe contributes only drag (at least from an aerodynamicist's point of view), why not build an aircraft that only has wing surfaces—a flying wing?

That is what Northrop and the Air Force did with designs such as the piston-powered XB-35 and jet-powered YB-49 of the 1940s. Jack Northrop conceived the flying wing, determining that if the entire weight of the aircraft was spread along the wingspan, the aircraft would be more aerodynamically efficient. With a flying wing, the weight is where the lift is, and the large bending forces found in conventional aircraft are eliminated. The wing, therefore, can have a very large span. The flying wing also eliminates much of the parasitic drag generated by a fuselage.

Although both flying-wing bomber designs flew successfully, neither went into production. The YB-49 was slower than the Boeing B-47 (which the Air Force eventually purchased), and without a major redesign, it could not carry the large nuclear weapons in use at the time.

Another difficulty was controlling the flying wing in flight. Limited tail control, due to the lack of the typical tail surfaces located at the end of a long fuselage, severely compromised lateral, longitudinal, and directional stability. The control systems of the day were not advanced enough to compensate for the limited tail control. Today, computerized stability-augmentation systems are up to the task. Because of both their structural efficiency and reduced drag, flying wings could be of interest in the future, especially in very large cargo transports where huge quantities of cargo could be carried in the wings (Fig. 3-1).

Fig. 3-1. Because of their reduced drag and bending loads, the flying wing could reappear in very large air freighters of the twenty-first century. A Boeing 747 is shown for comparison.

The flying wing is also of interest to the military in its quest for stealth aircraft with "low observables." At least when viewed from the front and sides, the flying wing has a small radar cross-section (RCS), the characteristic that determines how difficult it is for enemies to detect the aircraft on their radar screens. This is a result of the flying wing's flat profile. It lacks large vertical surfaces that reflect radar signals. Also, the engines, which give off detectable heat (infrared radiation) and can be destroyed by heat-seeking missles, can be buried deep in the wings. While no one will argue that aircraft like the 180 ton B-2 with its 170 foot wingspan is hardly "invisible" when it passes overhead, tests have convincingly shown that the B-2 cannot be consistently detected, tracked, or engaged at ranges that would be used on typical military missions (Fig. 3-2).

Northrop Corporation.

Fig. 3-2. The B-2 Stealth Bomber uses a flying wing configuration to achieve its low observable characteristics.

Swept-back and swept-forward wings

As the speed of the air over a wing increases to transonic and supersonic speeds, there is a dramatic increase in drag that can reach a point where the aircraft can no longer accelerate. To overcome this, wings have been swept back to ever increasing angles, as seen on fast-flying aircraft like the F-16, F-22, and Concorde (Fig. 3-3).

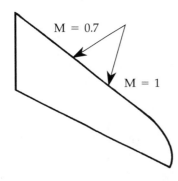

Fig. 3-3. When a swept wing is flying at supersonic speeds, it only "sees" that portion of the velocity that is perpendicular to the leading edge of the wing. Here, while the wing has a forward speed of Mach 1, because the wing is swept back 45°, the wing "thinks" it is only flying at Mach 0.7.

There are other advantages to sweeping the wings forward, even though such a design might appear weird at first glance. Like so many other ideas, forward-swept wings are not a brand-new idea. The Germans built the Ju-287 bomber with a forward-swept wing during World War II, and the German-designed HFB Hansa 320 business jet first flew with a forward-swept wing in 1964. The main motivation for these designs was the fact that by using forward-swept wings their attachment point could be moved farther aft. This permitted a larger bomb bay near the center of gravity, in the case of the Ju-287, and a longer cabin unimpeded by the wing span for the HFB business jet. The designers of the Ju-287 also realized that the forward-swept wing was a means of avoiding the classic problem of tip-stall found on rearward-swept wings. Tip-stall results in reduced lift and could even lead to pitching up or loss of lateral control as a stall condition is approached (Fig. 3-4).

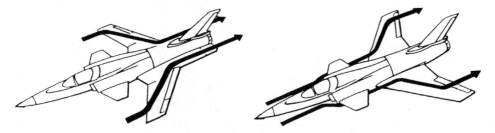

Fig. 3-4. With the forward-swept wing, the airflow tends to flow inward towards the wing root, while on the rearward-swept wing just the opposite occurs. This allows the former to remain unstalled up to a higher angle of attack. In addition, the stall tends to occur nearer to the wing root and, thus, is easier to control. Grumman Corporation.

The problem inherent in forward-swept wings that has prevented their widespread use until now is the fact that they become severely loaded at high speeds—the exact conditions where wing sweep is needed. A conventional wing washes out toward the tip; that is, the angle of incidence (the angle between the wing chord and the line of thrust) decreases toward the wing tip as it bends under lift loads. With a forward-swept wing just the opposite happens. The wing washes in, adding to the load at the tip and causing the wing to bend even more. In fact, at high speeds the loads can become so great that the results can be catastrophic. This phenomenon, called torsional or structural divergence, can be avoided by simply beefing-up the wing so it is able to withstand the large bending loads. But this results in an excessively heavy airplane if a conventional metal structure is used.

Fortunately, lightweight composite materials have come to the rescue of the forward-swept wing, so the level of interest in the design is currently high. Aero-elastically-tailored composites, as they are called, can be manufactured so that their fibers provide the greatest strength in the direction needed to resist the torsional divergence load, and they can do this without a weight penalty (Fig. 3-5).

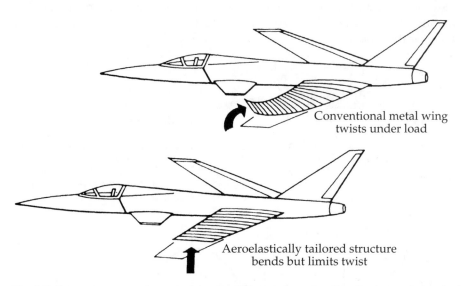

Conventional metal wing
twists under load

Aeroelastically tailored structure
bends but limits twist

Fig. 3-5. Because composite materials are stiff in only one direction, the forward-swept wing can be tailored to bend in a specific direction under aerodynamic loads, such as those that cause divergence. By crisscrossing the composite materials during construction of the wing, it can be designed to bend but resist twisting under the high loads of maneuvering. Grumman Corporation.

What are the other advantages of the swept-forward wing that makes it so attractive today? The swept-forward wing results in higher lift coefficients than its swept-back counterpart. While this factor might not be greatly appreciated during normal cruise, it would be during high-speed conditions such as a high-performance fighter would experience. Also, forward-swept wings do not have to be swept as severely to handle the drag-rise problem for the same Mach number. Because the leading edge has less sweep, a higher lift-to-drag ratio can be achieved.

The forward-swept wing can be shorter and still provide the same lift surface, shock sweep angle, and effective wingspan, resulting in a lighter wing. Conversely, by keeping the strength to resist bending loads and the structural weight the same, the wing's aspect ratio can be reduced, leading to less drag, which is always an important gain.

Supercritical wing

Even though an aircraft might not be flying at supersonic speeds, it is possible that the airflow over the aircraft, especially over the wings, can reach the point where it reaches supersonic speeds and there is a sudden and dramatic increase in drag. The airspeed at which this occurs is called the critical Mach number (Figs. 3-6 and 3-7). On a typical wing, the location where the airflow speed exceeds Mach 1 is usually near the wing's midpoint, or in aerodynamic terms, at about the 40-percent chord location. Here the typical wing has its greatest curvature and the air is moving at its fastest velocity. The critical Mach number is higher for thin wings than for thick ones.

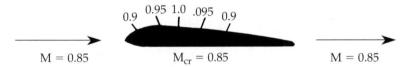

$M = 0.85$ $M_{cr} = 0.85$ $M = 0.85$

Fig. 3-6. The reason a wing provides lift is because the airflow across the upper surface is faster (and the pressure is lower) than the flow across the lower surface. Thus, even when a wing flies at supersonic speed, it is possible that the upper airflow will reach Mach 1 or even higher velocities. The flight Mach number at which the airflow is accelerated to exactly Mach 1 somewhere on the airfoil is called the critical Mach number. In this example, the aircraft is only at Mach 0.85 when the airflow reaches Mach 1.

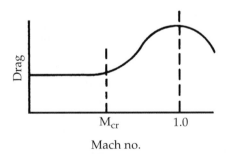

Fig. 3-7. At low subsonic speeds, the drag produced by an airfoil is essentially constant. As the freestream velocity approaches the critical Mach number, drag increases dramatically. Aerodynamicists call this *drag divergence*, and the critical Mach number is sometimes called the *divergence Mach number*. Drag continues to rise rapidly until about Mach 1. In the early days of flight, some experts thought drag would increase infinitely at Mach 1, and thus, the idea of the sonic barrier was in vogue. However, Chuck Yeager and his X-1 showed that the sonic barrier could be penetrated and that after Mach 1, drag starts to decrease again.

In the mid-1960s, Dr. Richard Whitcomb of NASA began his research on the supercritical wing. The supercritical wing has a special design that delays formation of a shock wave until a higher critical speed. The resultant drag rise is also delayed. Surprisingly, the supercritical wing is considerably thicker for the same lift-to-drag ratio. Thus, the designer can either keep the same thickness and reduce the drag, or maintain the same drag and have a thicker wing (Fig. 3-8). Actually, supercritical wings not only are of use on high-speed aircraft, but also are of benefit to commercial and general aviation aircraft, which fly much slower.

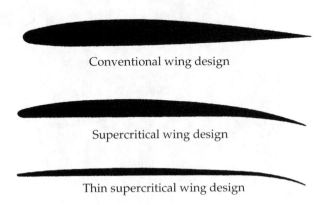

Fig. 3-8. The supercritical wing delays and reduces the strength of the shock waves on the upper surface at transonic speeds. The Grumman X-29 will be the first aircraft to fly with the thin supercritical wing. Grumman Corporation.

Conventional wing design

Supercritical wing design

Thin supercritical wing design

Winglets

Among the other important contributions made by NASA's Dr. Whitcomb are the winglets (first used on the Learjet) that are already seen on some of the latest aircraft and will be an important feature on aircraft of the future (Fig. 3-9).

An airfoil produces lift because of a difference in pressure on the upper and lower surfaces. At the tips of the wings this pressure differential causes air to flow from the lower to the upper surface. This results in a swirling flow of air that trails behind the aircraft, a phenomenon known as wing-tip vortices. The more lift a wing is producing, the greater is the strength of the vortices. Not only does the wing-tip vortex effect increase drag and decrease the lift produced by a wing, it can cause havoc around airports. The vortices shed by a heavy aircraft like a Boeing 747 can trail the aircraft for miles before they finally are dissipated. Other aircraft flying into the vortices can be severely buffeted. Flying light aircraft too close to an aircraft shedding vortices can be extremely dangerous.

Dr. Whitcomb came up with the winglet, a remarkably simple solution to eliminate the wing-tip vortices, or at least to reduce their strength substantially. A winglet added at a precisely determined angle with respect to the main wing not only overcomes the vortex problem, but improves the efficiency of the wing, leading to reduced fuel consumption. The wing-tip tanks found on many military fighters provide a similar effect, although not an optimum one, as in the case of a well designed winglet.

Fig. 3-9. The Learjet Longhorn 28/29 series was the first production aircraft in the world with winglets.

Wing-tip vortex turbines

NASA has taken the concept of the winglets one step further. In place of the winglets, wind turbines are added at the end of each wing. Besides reducing the vortex strength and drag, these turbines extract energy from the swirling air (Fig. 3-10). The turbines can be connected to generators that produce electricity to be used aboard the aircraft. Engineers predict that up to 400 horsepower could be generated if a similar system is installed on a large aircraft like a Boeing 747. The concept was successfully demonstrated in flight tests with a Piper PA-28 fitted with wingtip turbines on the trailing edge of both wings. Now it is up to the aircraft design community to use this off-the-shelf technology.

Mission-adaptive wings

A wing design is a compromise. For example, if a designer wants a wing for an aircraft that will cruise at supersonic speeds, he must also design the wing to fly at subsonic speeds as well as for takeoffs and landings. Thus, the optimum cruise configuration is compromised to gain the other necessary characteristics. Granted, devices such as ailerons, flaps, spoilers, and slats can partially compensate for deficiencies, but these still do not give the optimum performance of a wing designed for a particular flight regime.

Fig. 3-10. Tests of wing tip wind turbines installed on a Piper PA-28 showed that over twenty percent of the energy required to overcome drag induced by wing-tip vortices could be recovered.

The answer is a mission-adaptive wing that maintains its peak aerodynamic efficiency under almost all flight conditions (Fig. 3-11). This is done by changing the shape of the wing (like a bird does). Using flexible composite materials and an on-board computer for assistance, actuators buried below the surface of the wing can vary the surface contour as demanded by pilot control and flight conditions. Unlike flaps, spoilers, slats, and ailerons, which are appendages to the wing, the mission-adaptive wing is one piece (Fig. 3-12).

The oblique wing

As mentioned previously, swept-back wings are needed for high-speed supersonic flight. However, the swept wing does not provide enough lift to allow sufficiently slow approach and landing speeds. One way to have both good low-speed and high-speed flight characteristics in the same aircraft is to use a variable-geometry wing. For example, the main wings on the Air Force's F-111 are extended for takeoffs, landings, and low-speed flight, but they are swept back for flight at supersonic speeds. While this has been quite a satisfactory solution, the bearings and

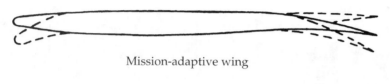

Mission-adaptive wing

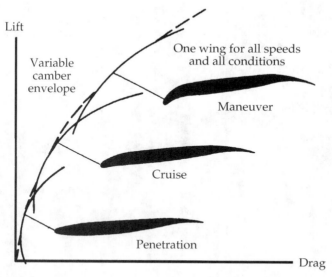

Lift

Variable camber envelope

One wing for all speeds and all conditions

Maneuver

Cruise

Penetration

Drag

Fig. 3-11. The mission-adaptive wing (MAW) provides optimum performance under varying flight conditions. This graph shows how well a MAW compares with an ideal variable-camber-envelope airfoil. The MAW automatically reconfigures during flight. Boeing Aircraft.

Boeing Aircraft.

Fig. 3-12. A prototype of the MAW was flight tested on the AFTI F-111.

mechanical parts for pivoting the wing must be very substantial, and thus quite heavy, in order to handle the bending loads produced by the wings.

A few year ago, prominent NASA aerodynamicist Dr. Bob Jones came up with a solution to the problem. (Incidentally, Dr. Jones was the instigator of swept-back wings in the 1940s.) The solution was the scissor, or oblique, wing. The oblique wing is a slender wing that can be pivoted about its center point (Fig. 3-13). For takeoffs, landings, and low-speed flight, the wing would be positioned at right angles to the direction of flight. For flight at supersonic speeds the wing would be pivoted to up to 65 to 70 degrees. Actually, the engines would do the rotating so that they are always pointed in the direction of flight.

NASA/Ames.

Fig. 3-13. The oblique airliner is the ultimate flying wing concept. Direction of flight would be determined by rotating engine nacelles and vertical fins at the ends of the wing.

The oblique wing aircraft would provide the best flight efficiency and fuel economy because it could be positioned for the optimum lift-to-drag ratios. For example, at Mach 1.4, about 1000 MPH, the wing would be positioned at 60 degrees, then the angle would be increased to the full 70 degrees for a Mach 2 flight (1400 MPH). Unlike the Concorde Supersonic Transport which is quite inefficient below the speed of sound, about 700 MPH, an oblique flying wing would fly at the most efficient sweep angle under all flying conditions.

While the oblique wing aircraft (Fig. 3-14) may look odd—one wing is pointed forward and the other one is pointed aft—it has very little effect on the aircraft's flight. Like a forward-swept wing, the wing itself does not care whether the sweep is forward or to the rear. The optimum wing platform shape for supersonic flight

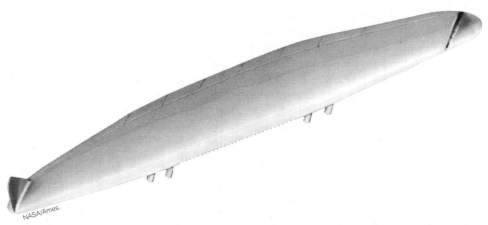

NASA/Ames.

Fig. 3-14. Flight tests of the AD-1 have successfully shown that the scissor wing will work.

is actually the same as for subsonic flight except for adjusting the wing's sweep angle depending on the flight speed.

The oblique wing could be used on an all-wing, or flying wing aircraft. The passengers and cargo would be carried within the wing itself. This flying wing design could benefit from technologies like computer-aided flight controls developed for the B-2 flying wing bomber.

The joined wing

Another concept that looks somewhat weird, but works very well, is the joined wing. Like other concepts, joined wings are almost as old as the airplane itself. However, the most recent interest in the concept has been advocated by Dr. Julian Wolkovitch, who holds most of the key patents on this unique design. The joined-wing aircraft has its main and tail wings joined near their tips so that, when viewed from either the top or front, the wings form a diamond (Fig. 3-15). To do this, the main wing is swept rearward and the tail wing is swept forward. The wings are joined at the tips, or at a point inboard of the front wingtips.

The advantages of the joined wing over a conventional wing include lighter weight, greater stiffness, lower drag, and higher lift, plus improved stability and control. The joined wing can also provide unique maneuvering capability. The lighter weight comes from the way in which the wings are constructed. For a conventional wing, a box-type structure is used to resist the bending loads that come primarily from lift. A large amount of material is needed on the top and bottom of the box structure, which means a lot of weight. With a joined wing, the lift load is broken into two components. One component is along the plane of the joined wing. The other, much larger, component is perpendicular to the plane formed by the wings. Because of the orientation of this perpendicular component of lift, the bulk of the wing material can be located near the leading and trailing edges. Because resistance to bending depends both on the amount of material and the beam

Fig. 3-15. The joined-wing concept could result in not only a lighter aircraft, but also a more maneuverable one.

depth involved, the large effective beam depth of the joined wing makes up for a lot of material, thus substantially reducing weight. The material in the leading and trailing edges is at just the right location to maintain leading edge contours for good aerodynamics and, at the trailing edge, to handle the loads produced by control surfaces and flaps.

Even though joined wings are thinner, they are stiffer than a conventional wing because torsional loads on one set of wings are resisted by the flexure of the other set. In wind tunnel tests, joined wings have been found to have less induced drag than conventional wings, as well as maximum lift coefficients. Thinner wings can be used, and because there is less wetted area (area in contact with the air), parasitic drag is reduced.

By putting control surfaces on both sets of wings, some interesting maneuvering is possible. By deflecting the front and rear surfaces in opposite directions, large pitching movements result. Deflecting all the surfaces down makes them act as flaps, providing high, direct lift control. To obtain powerful roll control, the surfaces can be used as ailerons. If the control surfaces on the front and rear wings are deflected to give equal, but opposite rolling movements, direct side-force control is obtained, and you have the unique ability to move sideways.

Canards

While canards are as old as the airplane itself, it has been only recently that they have come into their own. They will be a design focus of many aircraft of the future, from fighters to ultralights (Fig. 3-16).

Fig. 3-16. Burt Rutan pioneered both the canard and composite construction. Here they are used in their simplest form, a glider.

To prevent an aircraft from pitching in an oscillating manner, its aerodynamic center must be well separated from the center of gravity. For stability, the center of gravity is located forward of the aerodynamic center. In the past, most aircraft used horizontal stabilizers to counteract the nose-down attitude attributed to the center of gravity's forward location. But canards can perform this function while also producing positive lift, allowing the main wings to be smaller and lighter. By contrast, the typical horizontal tail produces negative lift, so the main wings must be larger to compensate for the loss of lift. In addition, canards are never in the downwash of the wings, so unlike a horizontal tail, they are effective at high angles of attack. This last characteristic is important in high-performance aircraft, which operate at very high angles of attack to achieve supermaneuverability.

Canards are now practical because aeronautical engineers now know how to prevent the instability problems experienced in the past. Most of the success in using canards can be traced to the work of aeronautical genius Burt Rutan.

Laminar flow control

Close to the surface of every wing in flight is a very thin sheet of air called the *boundary layer*. At low speeds, this boundary layer follows the contours of the airfoil and is very smooth. Indeed, under these conditions the flow is called *laminar* because the fluid particles are moving in well-ordered "layers." As the airspeed increases, the flow transitions to turbulent flow conditions where there is much mixing of particles among layers. Turbulent flow results in much greater surface friction and drag. The idea behind laminar flow control is to maintain laminar flow over the entire surface of the wing at high speeds, and thus reduce drag.

One method of obtaining laminar flow control is to suck off a portion of the boundary through holes or slots in the surface of the airfoil, or to construct the actual airfoil surface of a porous material (Figs. 3-17 and 3-18). Pumps below the airfoil surface are needed for the suction, and ducting is required to vent the air from the airfoil surface to the atmosphere outside.

Normal surface layer is thick and turbulent with high drag

Suction-stabilized surface layer is thin and laminar with low drag

Fig. 3-17. This is how laminar flow is maintained through suction of the boundary layer. NASA.

While the benefits of laminar flow (or *boundary-layer control*, as it is sometimes called) have been known for years, and even demonstrated in flight tests, the concept has not actually been put into application. For example, it offers over a 25 percent potential reduction in fuel consumption when used in subsonic commercial airliners. However, the pumps and ducting become quite complex and represent a sizable weight penalty. Also, means have not yet been perfected to prevent dirt and insects from clogging and corroding the very small suction holes and slots and the pumping and ducting systems. This currently presents an unacceptable maintenance problem, especially for commercial airline operators.

A much simpler approach to maintaining laminar flow is called *natural laminar flow* (NLF). By properly shaping the wing and making its surface extremely smooth, the transition from laminar to turbulent flow can be delayed so that laminar flow is retained over most of the wing. High technology has made NLF possible. The very accurate wing contour is designed by computer; the very smooth

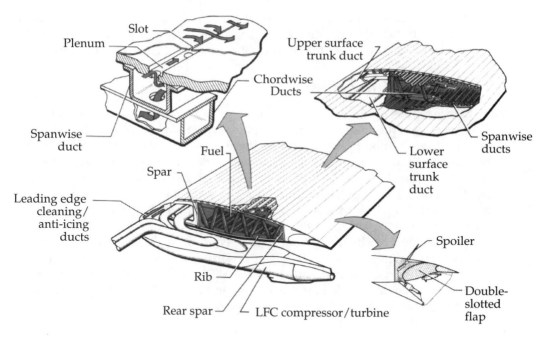

Fig. 3-18. Laminar flow control is considerably complex. NASA.

surface is a result of composite materials. Unfortunately, surface roughening by bugs and ice can still present a problem.

Circulation control

Somewhat akin to the idea of a boundary-layer control is the idea of circulation control. Here, however, air is blown out through one or more slots in a wing (rather than being sucked into the wing) to increase the circulation of the free-stream air around the airfoil and thus substantially increase the lift.

Typically, a source of compressed air, usually an engine-driven pump or bleed air from the turbine, supplies pressurized air to a plenum chamber in the wing. The air then flows through a slot located near the trailing edge of the wing. (Some designs also call for slots in the leading edge.) This escaping air causes the air flowing over the top of the wing to accelerate. By accelerating the air, lift is increased. Drag, however, is decreased.

The X-wing

The X-wing carries the concept of the oblique or scissor wing one step further. The X-wing operates like a helicopter with the wing serving as rotor blades during low-speed flight. Once the "rotors" accelerate the aircraft to around 200 knots (about the top speed for a conventional helicopter), the wing is stopped and locked into place. At high speeds, the blades function like conventional wings.

Circulation control, mentioned previously, would probably be an integral part of the X-wing aircraft. To prevent a helicopter from rolling over, the amount of lift produced by the rotor blades changes constantly as the rotor rotates. Also, the lift of the blades is changed as the helicopter climbs, descends, turns, hovers, or increases or decreases its forward speed. In a conventional helicopter, the change in lift is accomplished by changing the pitch (angle of attack) of the blades. This requires a complex arrangement of mechanisms that includes a swash plate, pitch horns, control rods, levers, and other mechanical paraphernalia.

With a circulation-control rotor design, changes in lift would be accomplished simply by regulating the amount of air blown through the slots (Fig. 3-19.) The rotor, which would be rigidly attached to the mast, would still rotate mechanically through a transmission connected to the engine. A set of valves would be used to control the amount of compressed air passing through the plenum in the wings and out the slots. The valves would be controlled by the pilot as part of the flight control system.

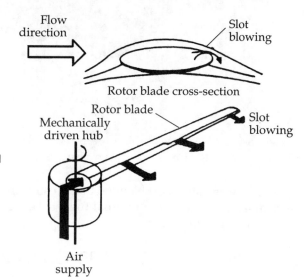

Fig. 3-19. Circulation control would enhance and control lift on a helicopter rotor blade. The concept could also be used on a conventional wing.

Not only does the circulation-control rotor offer the potential for a simpler, less expensive, and more easily maintained rotor control system, it operates more quietly and with less vibration. This is important in making helicopters more comfortable, more reliable, and in the case of military helicopters, less observable.

Lifting bodies and blended bodies

High-speed aircraft of the future will use fuselages that provide lift, resembling the Space Shuttle's lifting-body design. For aircraft, the design is commonly referred to as a "blended" body (Fig. 3-20), which is especially effective in providing lift at high angles of attack (where the lift from conventional wings starts to

Fig. 3-20. The F-16's blended-body is clearly seen here.

drop off). In addition, the blended body results in less drag (due to less wetted surface area), provides more internal fuel-storage capacity, and allows a more rigid and lighter-weight structure.

FLIGHT CONTROLS

There is more to aerodynamics than the shape of wings, fuselages, and empennages. There are the various control surfaces, such as elevators, rudders, flaps, and so forth. How these devices are controlled has a significant effect on the aircraft's performance characteristics.

Flight controls that are now on the drawing board, being tested in the laboratory, and being flown on research aircraft will allow aircraft to do some amazing things and, in the process, will result in better fuel economy, longer aircraft life, less maintenance, safer flying, and even a better ride in turbulent air.

Active flight controls

In a conventional aircraft, the pilot uses his stick (or control yoke) and his pedals to turn, climb, or descend. Active controls are automatic controls that operate independently of the pilot. A computer receives input commands from sensors located at appropriate locations on the airplane, makes its decision, and transmits this decision in the form of electronic commands to the control surfaces. The pilot still commands the aircraft, but active controls respond to external forces that happen so rapidly that the pilot can't take action fast enough.

One example of how active controls might operate on a commercial transport is in the case of turbulence.

Turbulence is something airline pilots like to avoid, not only because passengers dislike it, but because it puts a tremendous strain on the airframe. Circumstances such as bad weather or air traffic, however, might require airliners to fly through bad turbulence.

Suppose a turbulence sensor was mounted on a boom extending out in front of the airliner. The sensor would feel the turbulence ahead of the airplane and also determine its direction and strength. This information would be sent to the control system, where it would be translated into commands for corrective control action. All of this would happen in a fraction of a second, much faster than a human could sense the turbulence and react. Not only would such a system result in a smoother ride, it would greatly reduce the bouncing and pounding an airplane takes when flying in turbulence. The airframe would last longer and it could even be much lighter, resulting in reduced cost.

The military could also use turbulence sensors for its low-flying bombers. To avoid radar detection, bombers fly "on-the-deck" at high speeds with severe turbulence. Turbulence hammers the bomber's airframe and bounces the crew around so much that they cannot focus their eyes on their instruments or the world outside.

Active controls could also be used to change the lift distribution across the span of the wing. The lift produced near the tips could be reduced and the lift produced near the roots (where the wing joins the fuselage) could be increased. This would result in reduced bending loads, and wings could be made lighter.

Because wings are so flexible, they can vibrate rapidly at high speeds and when maneuvered violently, causing what is commonly called *flutter*. Wing flutter can become so severe that the wings can be torn off the fuselage. The normal approach to solving flutter problems is to make the wing stiffer, and usually heavier. An active control system could sense the onset of flutter and automatically adjust auxiliary control surfaces on the wing to increase the stiffness "aerodynamically" with a much smaller weight penalty.

Finally, an aircraft is highly stressed during maneuvers. For example, unlike straight-and-level flight (where loads might be large but are symmetrical), in a sharp turn, the wing can be bent or twisted in a nonuniform manner. The wing must be overdesigned to handle this nonsymmetrical loading. By using a form of active control that automatically deflects the trailing edge of the flaps, the lift distribution can be tailored to reduce wing bending.

Aeroelastically tailored wings

As mentioned earlier, composite material woven into wings has made the high-speed forward-swept-wing aircraft feasible. This material also offers a form of "active" control that does not involve the complexity of sensors or computers.

Composite wings can take advantage of the unidirectional stiffness of their materials. Wings made of composites consist of layers, or plies, of material woven on top of one another. By carefully designing the way the plies are woven (i.e., the direction in which the strongest fibers run), the wing can be made to bend so as to give the best lift and drag characteristics for a given set of flight conditions.

Relaxed static stability

Conventional aircraft have been designed to automatically return to straight and level flight after being disturbed by a gust or a pilot command. This is called *positive static stability* and is the result of the proper design and location of the horizontal and vertical tails as well as wing dihedral.

Although positive static stability might be desirable in a "forgiving" light airplane or even a commercial airliner, it is definitely a hindrance to superior maneuverability in a high-performance fighter. Typically, fighters are designed with little or no static stability.

While providing good maneuvering characteristics, relaxed static stability requires the pilot to fly the aircraft, hands-on, all the time. To reduce pilot workload, and in some cases even make the aircraft possible to fly in the first place, a stability augmentation system is needed to compensate for the lack of static stability—a form of active control.

Fly-by-wire control systems

In the early days of aviation, flight control surfaces were connected to the pilot's controls through a maze of cables, bell cranks, pulleys, and a variety of other mechanical components. In later years, these mechanical systems were augmented by hydraulic control systems to enhance the pilot's control in high-performance military aircraft and commercial transports. These systems were very complex and added significantly to the weight of the aircraft.

With the recent revolution of fly-by-wire systems, the mechanical connections between a pilot's controls and the control-surface actuators are replaced by wires that carry electrical signals. In addition to the wires, various electromechanical devices are used that convert stick and rudder-pedal motions into electrical voltages. These voltages are measured, and their values are fed into a digital computer. The computer is programmed with a set of control laws that make the aircraft "flyable." The computer output, again in the form of electrical signals, is fed by wire to the actuators, which, in turn, move the control surfaces.

Fly-by-wire control systems were first used on early manned spacecraft like the Mercury, Gemini, and Apollo satellites. Much more sophisticated systems were installed on the Space Shuttle. Military fighters like the General Dynamics F-16 and Lockheed F-22 already use fly-by-wire control systems. Fly-by-wire is also found on the Concorde and Airbus Industrie A320 commercial airliners.

Besides reducing weight, fly-by-wire provides more responsive controls and makes multiple, redundant, flight-control systems easier to build. Also, it is possible to change the flight characteristics of an aircraft simply by reprogramming the control laws in the computer. In the future, it might be possible to change an interceptor to a ground-attack fighter or a photo-reconnaissance aircraft by simply rewriting computer software and changing weapons and mission equipment. The Swedes applied the concept in the Saab JAS-39 Gripen multirole combat fighter.

Another interesting example of fly-by-wire technology is the self-repairing flight-control system being developed by the Air Force. In combat, an aircraft's primary control surfaces could suffer battle damage or even be completely blown away. By instantaneously reconfiguring the flight-control system, primarily through alternate flight-control laws, the surviving control surfaces (like rudders, flaperons, ailerons, or stabilizers) could be used in combination to perform the functions of the lost surface. This would not only keep the aircraft from crashing, but also allow the aircraft to fly safely home and, in many cases, even complete its mission.

Fly-by-light control systems

In fly-by-light systems, optical fiber replaces wires. Signals are transmitted along the optical fibers at the speed of light using light from a laser. Because a single optical fiber the thickness of a human hair can carry a tremendous amount of information, heavy wiring can be replaced with featherweight optical-fiber cables. Fiber optics are also impervious to electromagnetic radiation, an important consideration for military aircraft that must operate in an environment of electronic countermeasures. Fly-by-light control systems can be built without electromagnetic shielding, further reducing total aircraft weight.

AERODYNAMICS AND SURVIVABILITY

A military aircraft's survivability is enhanced by being able to escape enemy detection. One of the biggest challenges facing military aircraft designers today and in the future is the need for aircraft with "low observables," that is, aircraft that are "invisible" to sophisticated enemy sensors. These sensors can detect an aircraft by its radar, infrared, acoustic, and visible signatures. Indeed, a sensor can be developed to discriminate the signature of any airplane characteristic, and it can be used to detect and destroy the plane.

The other part of the survivability equation is the ability of the pilot to survive enemy interceptors, guns, and missiles. A key way to accomplish this is by maneuvering to avoid them. Aerodynamics play the significant role here, and today there is a great deal of research being done to provide aircraft, primarily fighters, with super-maneuverability.

Supermaneuverability

Supermaneuverability is defined as the capability of a fighter to perform maneuvers at angles of attack beyond the point of stall, as well as controlled side-slip maneuvers. The need for supermaneuverability has been brought about by increased capabilities of potential enemy weapons systems.

One of the first operational aircraft to use a form of supermaneuverability was the AV-8 Harrier V/STOL fighter with its vectored-thrust engines. Normally, the exhaust of the engine points downward for vertical flight and rearward for horizontal flight. However, by pointing the nozzles a bit forward, the harrier can perform some maneuvers that are guaranteed to confuse (and hopefully defeat) the enemy. Using vectoring-in-forward flight (VIFF), the Harrier can decelerate more rapidly than any other aircraft. In a dogfight with the enemy on its tail, the Harrier can slow down so suddenly that, to the adversary, it appears to have stopped in midair. The hunter would then be the hunted. This is just one of a multitude of interesting tactics available with VIFF.

A few years ago, the Air Force modified an F-16 into the Controlled Configuration Vehicle (CCV-16). The main changes were the 8 square-feet ventral canards mounted on each side of the engine inlet duct (Fig. 3-21). To move the aircraft's center of gravity farther back, changes to the fuel tank arrangement and to the flight-control system were made. These changes radically changed the aircraft's flying characteristics. For example, the aircraft's nose could be pointed in any direction without changing the direction of flight. It could move sideways without banking or rolling. It could also climb, descend, or move sideways without changing the direction of the nose. Just think of the possibilities this would give the fighter pilot.

General Dynamics.

Fig. 3-21. Ventral canards on a CCV-modified F-16 allowed it to perform some radical flight maneuvers.

Basic research has been done on wing aerodynamics at angles of attack beyond the point of stall. While wing stall has always been something designers and pilots have religiously tried to avoid, in the future aircraft could fly in the post-stall regime to achieve supermaneuverability.

For example, wind tunnel research has shown that by rapidly pitching an airfoil past the stall angle of attack, it is possible to achieve several times the lift compared to the same wing flown in normal flight. The results of the research, commonly categorized under the generic title "unsteady aerodynamics," offers some interesting possibilities for the future. Scientists are even studying in great detail how Mother Nature's creatures, such as the dragonfly, perform their unique flight maneuvers, hoping someday to adapt what they learn to manmade aircraft.

Future high-performance fighters will be able to do things only dreamt of by today's hottest fighter pilots.

AERODYNAMICS RESEARCH AND DEVELOPMENT

As the saying goes, "the proof is in the pudding." In the development of new aerodynamic concepts, the proof comes during the idea's flight test. Building manned experimental aircraft for flight testing is an expensive proposition, so many techniques have been developed to try out new ideas relatively inexpensively, saving final testing to the very end.

Wind tunnels

The Wright brothers were using wind tunnels to do aerodynamic research before their historic first flight at Kitty Hawk. Through the years, wind tunnels have grown in sophistication to keep up with the need to fly faster and higher. This growth will continue into the future with new capabilities derived from advances in electronics, materials, and propulsion technologies.

The Reynolds number is a key parameter in describing the capabilities of a wind tunnel. The Reynolds number is defined by:

$$\text{Reynolds number} = \frac{\text{velocity} \times \text{characteristic dimension}}{\text{fluid viscosity}}$$

The velocity is the freestream speed of the air flowing through the wind tunnel, or the airspeed of the actual airplane. The characteristic dimension could be the chord length of an airfoil or the diameter of an engine inlet duct. Because a scale model has a smaller characteristic dimension than an actual design, the tunnel speed must increase, the viscosity decrease, or both, if the same Reynolds number is to be attained. The idea is to have the Reynolds number in the test as close as possible to the Reynolds number of the actual aircraft. As flight speeds and aircraft size increase, this becomes much more difficult. For example, a giant air freighter might have a Reynolds number of around 125 million. Until recently, the best of NASA's wind tunnels could only simulate Reynolds numbers of about 15 million.

A major step forward was taken in the early 1980s with the opening of the National Transonic Facility, located at the NASA's Langley Research Facility in

Virginia. The development of aircraft that will fly at hypersonic speeds (Mach 5 to Mach 25) will place even greater demands on wind tunnels. In recent years there has been a substantial effort in the United States to upgrade tunnel capabilities to meet these needs (Fig. 3-22).

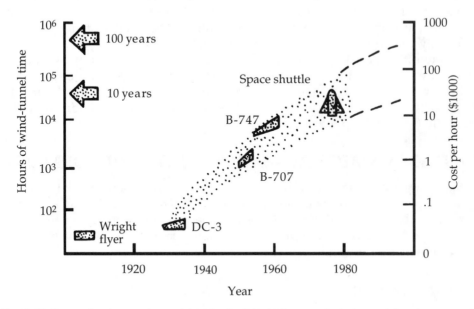

Fig. 3-22. From a few hours of tunnel time for the Wright flyer, total wind tunnel time for a new aircraft has increased to years. Cost per hour of tunnel time has also increased exponentially. Testing of the National Aerospace Plane will undoubtedly fall at the very top of this chart.

In addition to the high-technology tunnels themselves, great strides are being made in the methods used to take data and gain information during tests. Because wind-tunnel testing is very expensive, it is important that data acquisition be done very efficiently. High-speed computers are a major factor here, as is the laser.

The laser velocimeter has become a very important tool in obtaining data during wind-tunnel tests. Not only does it gather data at an extremely high rate, it does not disturb the airflow, unlike other flow measurement devices. Disturbed airflow sometimes causes erroneous results. The laser velocimeter and laser interferometer can also be used to look at the pattern of the airflow over the wind-tunnel model, replacing older techniques such as injecting smoke into the airstream or placing tufts of thread on the model. With the laser, aerodynamicists can get a moving picture of what is happening and a better understanding of the complete flow phenomenon (Fig. 3-23).

Other important technological advances are being made to improve the accuracy of wind-tunnel simulations. For example, in wind-tunnel testing, the model must be supported in some manner. Today, this is usually done by a "sting" attached to the rear of the model. The sting supports the model and contains connections so information from sensors imbedded in the model can be transmitted to

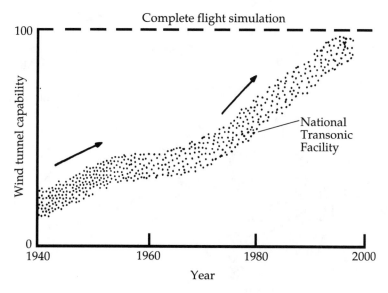

Fig. 3-23. Eventually, the full flight envelopes of many aircraft will be completely simulated in wind tunnels, eliminating most, but not all, expensive flight testing.

the instrumentation outside the tunnel. The sting, unfortunately, disturbs the airflow and adds extraneous readings that can lead to erroneous results.

Alternatively, the models could be suspended by magnetic forces. For example, in the Magnetic Suspension and Balance System (MSBS) at NASA's Langley Research Center, models weighing as much as 6 pounds are suspended in midstream using powerful electromagnets (Fig. 3-24). Not only do the magnetic forces suspend the models, they can measure the aerodynamic forces exerted on the models, which is information the aerodynamicist needs. By varying the strength of the magnetic field, it might even be possible to accelerate and maneuver the model to simulate real flight conditions. The model could actually be flown in the wind tunnel.

The electronic wind tunnel

The mathematics of aerodynamics are among the most complex of any physical science. Even a simple problem (e.g., solving the equations to describe non-turbulent airflow around an aircraft in straight-and-level flight) can take as many as 50 billion or more computer operations. When the flight is complicated by complex aircraft geometries, maneuvers, or turbulence, the number of operations jumps by orders of magnitude. Because of these seemingly insurmountable problems, aeronautical engineers and designers in the past relied on the results of wind-tunnel tests and simple approximate solutions to mathematical equations.

Because wind-tunnel testing has become so expensive and even minor changes in a design can require a whole new set of tests, a great investment is being made in supercomputers that can solve the complex mathematics of

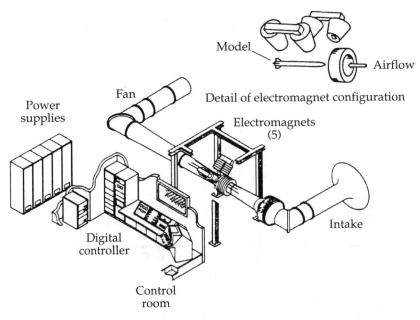

Model

Airflow

Detail of electromagnet configuration

Power
supplies

Fan

Electromagnets
(5)

Digital
controller

Intake

Control
room

Fig. 3-24. The Langley 13-inch Magnetic Suspension and Balance System (MSBS) is a laboratory established to develop technology required for wind-tunnel testing that is free of model-support interference. NASA/Langley.

aerodynamics (Fig. 3-25). Today, the hallmark of computer technology is the Cray Y-MP M90 with the largest memories ever offered. For example, the eight-processor M98 model provides 32 gigabytes of memory capacity.

The question is often asked as to whether the electronic wind tunnel will ever replace the real wind tunnel. The answer is a definite no. The two will always be around to complement one another. Experts say that computers able to design a complete aircraft, including its engines, will not arrive for another 25 years. Such a computer would require capabilities at least 1000 times more powerful than the supercomputers now emerging. And there will probably always be aircraft geometries and flight regimes that defy simulation by mathematical equations and computers. Also, no matter how accurate computer simulations become, wind-tunnel data will be required to verify the accuracy of the computer results. As the mathematical solutions are extrapolated to new regimes, test data is needed at various points along the way to ensure that the extrapolations are pointed in the right direction.

Flight tests

Aircraft designers must have confidence in a new aerodynamic concept before it is accepted and actually incorporated into production aircraft. Normally, this confidence is gained after a triad of validation techniques. The triad consists of mathematical/computer simulations, wind-tunnel testing, and flight testing. Be-

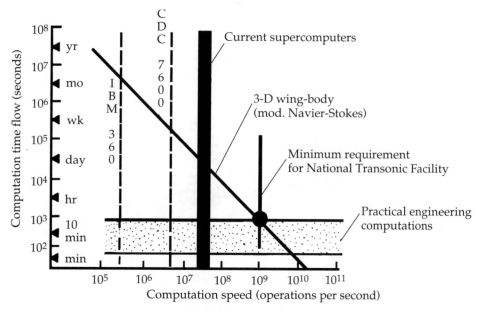

Fig. 3-25. Computer generated airflow patterns can tell much about the aerodynamics of future designs like the oblique wing airliner without the need for much more expensive wind tunnel testing. NASA/Ames.

cause flight testing is so expensive, it is usually saved until last. At that point, only one or two of the many designs tested have passed the computer and wind-tunnel tests and are actually installed on a test aircraft.

In the past few years, a couple of rather revolutionary techniques have emerged to reduce the costs of flight testing. One of these is the subscale flying model. The flying model is made of new composite materials and built with techniques borrowed from the light-aircraft and homebuilt-aircraft builder. The model is only a fraction of the size of the real airplane and can be built and tested for a fraction of the cost. Designs for such diverse aircraft as Beech's Starship I, the Air Force's Fairchild-developed T-46A trainer, and the oblique wing airplane have already been built and tested in subscale form (Figs. 3-26 and 3-27).

One of the greatest expenses involved in building flight-test aircraft is in their man-rating, that is, incorporating safety and life-support equipment for the pilot. Cost savings can now be realized by the use of sophisticated radio-controlled, unmanned models called *remotely piloted research vehicles* (RPRVs). RPRVs can be built inexpensively and tested in flight regimes that could prove hazardous in a human-piloted test aircraft. For example, future concepts for fighters with supermaneuverability were flight tested with the HiMat RPRV, including being air-dropped from a B-52 mothership, and flown through some very demanding maneuvers by radio control. Another advantage of RPRVs is that they can be easily modified to try out variations in a concept without the costly need to requalify the aircraft for manned flight.

Fig. 3-26. Scale model of the oblique wing concept being flight tested via radio control.

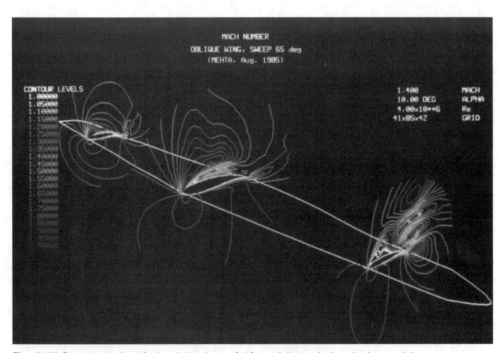

Fig. 3-27.Computer aircraft simulation is useful for solving aviation design problems.

4

Propulsion

THROUGHOUT the history of aviation, propulsion has been the technology that has made milestone flights possible—from the Wright brothers' first flight to the Space Shuttle. In some cases, the lack of suitable aircraft powerplants has prevented attractive concepts from coming to fruition. Propulsion will continue to play this commanding role in the future.

SETTING THE PACE FOR CHANGE

While the Wright brothers recognized the advantages of the internal-combustion engine, they found that the available automobile engines were much too heavy. So they designed and built their own 12-horsepower engine with a power-to-weight ratio of 1-to-15, which is pretty good for its day. Similarly, they found propeller technology was insufficient, so again they used their aeronautical knowledge and innate ability to design a very efficient propeller that provided a revolutionary advance in propeller technology.

Up until the jet age, the reciprocating engine was the most viable aircraft powerplant. Advances in internal-combustion engines came quite rapidly, with most aircraft using radial, inline, or V-type cylinder layouts that were either air-cooled or water-cooled. By the end of World War I, engine output had jumped to 400 horsepower in powerplants like the 12-cylinder Liberty engine, perhaps outpacing the fragile airframes they were attached to.

One of the major propulsion advances of the pre-World War II era was the supercharger. The supercharger compresses the thin air of high altitudes so engines can get sufficient amounts of air for proper combustion to produce sufficient power. It first allowed military aircraft, and then commercial airliners, to fly at altitudes high enough to avoid most adverse weather. The supercharger also produced an important side benefit, the pressurized cabin airliner. Both of these advances helped make commercial air travel an everyday occurrence, starting with the pioneering Boeing 247 and Douglas DC-3 (both with super-charged engines), and the Boeing 307 Stratoliner (the first pressurized-cabin airliner to go into commercial service).

Incidentally, while most engine manufacturers concentrated on superchargers driven by gears off the engine crankshaft, General Electric in the United States worked on perfecting the turbo-supercharger, or turbocharger. The turbocharger is powered by hot gases that pass through a turbine, extracting power from the exhaust that normally would just be wasted out the exhaust pipe.

By the early 1950s, the piston engine has reached its zenith with the Wright Turbo-Compound, an 18-cylinder engine that could produce up to 3700 horsepower. This engine allowed airliners like the Douglas DC-7, the Lockheed Super Constellation, and the Starliner to fly across the continent or the Atlantic Ocean nonstop. But jets were starting to replace the internal-combustion engine in high-performance military aircraft and large commercial airliner arenas.

By the end of World War II, the jet-powered Lockheed P-80 Shooting Star, the Air Force's first combat jet, had made its maiden flight, and the jet age was here. By 1949, the first jet airliner, the de Havilland Comet, was flying, but because of a series of unfortunate crashes traced to airframe structural fatigue, the Comet never received the fame and fortune it rightfully deserved. That would go to the Boeing 707, with its JT3C turbojet engines adapted from the jet engines used in such Air Force craft as the North American F-100 and Boeing B-52.

The turboprop, a spinoff of the pure turbojet, brought the jet age to commuter and business aircraft markets. The turboprop was more suited to these markets because of its high thrust at low to medium subsonic speeds and good fuel economy in comparison to the fuel-thirsty turbojet. The turboshaft engine, a close relative of the turboprop, replaced the complex, heavy, and maintenance-intensive reciprocating engines that had powered all helicopters until the Kaman K-255 first flew in 1951. Lightweight, with less vibration and noise, this small gas-turbine engine was popularized by the famous Bell Huey Helicopters.

COMBINED-CYCLE ENGINES

Our look into future propulsion systems begins with the application that will place the greatest demands on propulsion technology, flight in the hypersonic regimes—Mach 5 and beyond. What makes these applications especially challenging is the fact that not only must the aircraft fly at very high speeds and operate at very high temperatures, but it must also be able to fly slow enough to take off and land on conventional airport runways. And to keep weights reasonable, as well as to reduce complexity, it would be best to use a single engine-type called a *combined-cycle engine*.

For low-speed operation, the combined-cycle engine would operate like a normal turbojet. As the aircraft accelerates, it would operate like a ramjet, and then a scramjet. If the hypersonic aircraft is used for placing payloads into orbit, the combined-cycle engine could even be designed to operate as a rocket in space, where there is no oxygen. All of this would be done with a single engine, converting from cycle to cycle (thus the title combined cycle) by varying the configuration and geometry of the air passages through the engine. This does require a rather elaborate control system.

In a turbojet engine, the incoming airflow is slowed down, both to convert the velocity to pressure (to provide thrust) and so that the fuel-air mixture can be ignited and burned in the combustion chamber. A compressor is needed in front of the combustion chamber to further increase the pressure. A turbine section is located behind the combustion chamber to extract power from the exhaust gases to drive the compressor and various engine accessories. As flight speeds increase, the compressor and turbine can be eliminated to create a ramjet, in which the entire conversion of velocity to pressure is done by the ram effect without a need for any rotating machinery (Fig. 4-1).

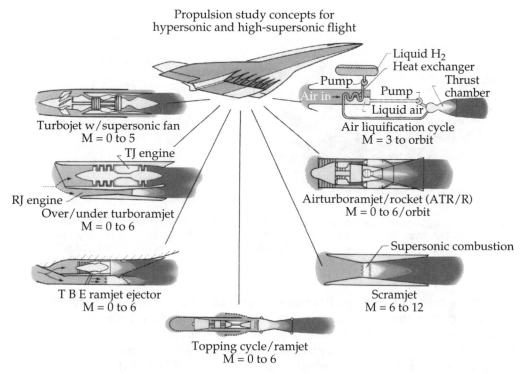

Propulsion study concepts for hypersonic and high-supersonic flight

Turbojet w/supersonic fan
M = 0 to 5

Liquid H$_2$
Heat exchanger
Pump
Thrust chamber
Air in
Pump
Liquid air

Air liquification cycle
M = 3 to orbit

TJ engine
RJ engine
Over/under turboramjet
M = 0 to 6

Airturboramjet/rocket (ATR/R)
M = 0 to 6/orbit

T B E ramjet ejector
M = 0 to 6

Supersonic combustion

Scramjet
M = 6 to 12

Topping cycle/ramjet
M = 0 to 6

Fig. 4-1. This shows how a combined-cycle engine would allow the National Aerospace Plane to take off, land, and fly at all speeds from low-subsonic to orbital velocities. NASA.

Unfortunately, as air is decelerated, its temperature rises. If a normal turbojet engine is used, the faster the airplane travels, the more the air has to be decelerated, and the higher its temperature becomes. The maximum speed of a turbojet is about Mach 4. Above these speeds, the temperature inside the engine is so hot that metals within the engine lose their strength and can melt. The ramjet with its few moving parts partially overcomes this problem, but it is also limited to top speeds of Mach 5 to 6. The problem is that the air must still be decelerated to low-subsonic speeds for combustion to take place.

In the scramjet the air can be mixed with fuel and ignited while still traveling at supersonic speeds. As a result, temperature increases and pressure losses due to shocks are greatly reduced. Because the scramjet works on the ramjet principle at supersonic speeds, the name scramjet, short for supersonic combustion ramjet, is quite fitting (Fig. 4-2).

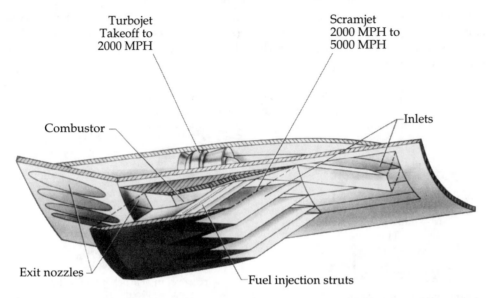

Turbojet
Takeoff to
2000 MPH

Scramjet
2000 MPH to
5000 MPH

Inlets

Combustor

Exit nozzles

Fuel injection struts

Fig. 4-2. This engine would operate like a turbojet at low speeds and a scramjet at high speeds. Unlike current engines that are often just attached to the aircraft's fuselage, the scramjet would be an integral part of the airframe, and the fuselage itself would be part of the inlet and nozzle. NASA.

The scramjet, however, has its own set of problems. One of these is the fact that there is very little time for ignition and combustion to take place, because the fuel-air mixture is moving so fast through the combustion zone. Hydrogen is probably the only fuel that has rapid enough ignition and combustion properties.

Even though the idea of the scramjet is quite simple and it essentially does not have any moving parts, it still represents a real technological challenge, one that engineers are only just beginning to get a handle on. Facilities are now available for use in the development of scramjets, at least up to Mach 8 (Fig. 4-3). The design and analysis of the scramjet also requires a tremendous amount of computer power, which fortunately is now available with the new generation of supercomputers (Fig. 4-4).

Like the ramjet, the scramjet is not capable of accelerating from zero velocity. Indeed, the scramjet cannot really start operating before speeds of about Mach 6. Other propulsion devices are needed to get the scramjet up to its operating speed, thus the reason for interest in the combined-cycle engine mentioned earlier.

The combined-cycle engine would be fully integrated into the aerospace vehicle it powers, not just a powerplant attached to the airframe. For example, in hy-

Fig. 4-3. A nozzle design for the National Aerospace Plane being tested in a NASA wind tunnel. During the tests, hydrogen fuel was burned in an external stream of air passing adjacent to the cowl section of the nozzle to reduce transonic nozzle drag.

personic designs, the forward part of the lower fuselage would actually be the engine intake while the rear of the fuselage would function as an exhaust nozzle.

Testing of scramjets at their operating speeds probably will not be done in wind tunnels, but in actual flight tests (Fig. 4-5). Several approaches for this flight testing are being proposed. The most ambitious and costly being the X-30 hypersonic test vehicle that would be manned, take off and land like an airplane and have the capability to reach orbital speeds.

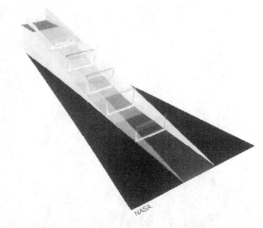

Fig. 4-4. Much of the research on high speed propulsion systems is being done on the supercomputer. Here computational fluid dynamics is used to predict the airflow velocities through a hypersonic engine inlet.

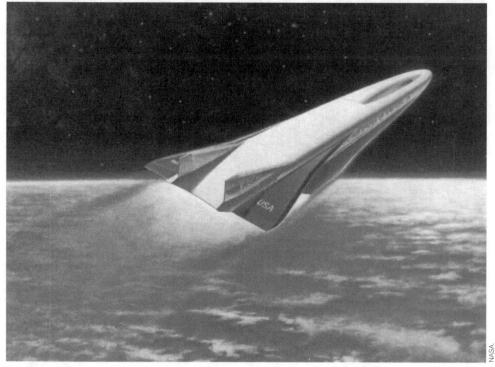

Fig. 4-5. The X-30 is a very ambitious and expensive approach to flight testing a scramjet as well as testing the materials and structures that could withstand the vigors of flight at speeds up to Mach 25.

Other alternatives include a manned HALO (Hypersonic Air Launch Option) test platform that would be carried to Mach 3 and a 70,000-foot altitude on top of a NASA/Lockheed SR-71. After launch, the HALO would accelerate via a rocket engine to about Mach 9 where scramjet testing would commence at speeds reaching up to Mach 12. Or an unmanned scramjet testbed could be launched on top of a rocket-propelled launch vehicle, such as a surplus Minuteman 2 ICBM for testing at speeds of up to Mach 25. It would then be operated for a short period to obtain vital test data that could be used to validate computer designs. Incidently, the Russians have already launched a scramjet test model atop one of its SA-5 missiles. Alternatively, testing might be done by using a much less expensive unmanned testbed primarily powered by scramjets and launched atop a missile or from another airplane. Its top speed could fall far short of the Mach 25 orbital speeds.

The variable-cycle engine, combining the features of both the turbofan and turbojet, has been proposed and developed by several aircraft engine manufacturers. The turbofan would be used for take-off through high subsonic speeds where it offers low fuel consumption, high thrust, and quiet operation (Fig. 4-6). For supersonic cruise, the engine would convert to the turbojet mode. The variable cycle could be used in future high-speed military fighters or a supersonic transport (Fig. 4-7).

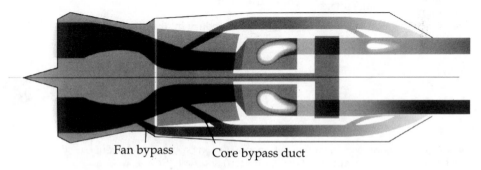

Fan bypass   Core bypass duct

Fig. 4-6. Simplified schematic drawing of airflow through a variable-cycle engine that can operate in either turbofan or turbojet modes. GE Aircraft Engines.

THRUST VECTORING

When it comes to revolutionary aeronautical innovations, thrust vectoring ranks up there with the jet engine, stealthiness, composite structures, and fly-by-wire controls. Until quite recently, piston-prop or turbojet engines were installed on aircraft strictly to provide thrust. Controlling the direction of flight was left up to the control surfaces like elevators, rudders, and so forth. Then came the AV-8 Harrier, originally developed by Hawker Siddeley (now British Aerospace) in England, and later built by McDonnell Douglas. The Harrier is in service with the U.S. Marine Corps, as well as in Britain and Spain (Fig. 4-8).

What made the Harrier a success was its vectored-thrust Rolls-Royce Pegasus engine. The idea behind this vectored-thrust engine is quite simple. The engine has four nozzles, two at the front and two at the rear, which use exhaust gases to produce thrust in the normal fashion. The pilot can "vector" the thrust by swiveling the nozzles downward for vertical thrust and rearward for forward flight. He can also set them at intermediate positions or even rotate them a bit forward of

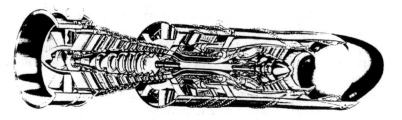

MCV 99, the SNECMA concept of the variable cycle engine.

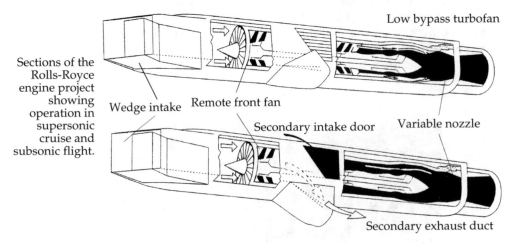

Low bypass turbofan

Sections of the Rolls-Royce engine project showing operation in supersonic cruise and subsonic flight.

Wedge intake   Remote front fan

Secondary intake door   Variable nozzle

Secondary exhaust duct

Fig. 4-7. A second generation supersonic transport would probably use a variable cycle engine. This illustration shows concepts from SNECMA (top) and Rolls-Royce. Aerospatiale.

British Aerospace.

Fig. 4-8. The AV-8 Harrier pioneered thrust vectoring. Besides allowing vertical takeoffs and landings, it provides some unique maneuvering capabilities when used in "vectoring in forward flight" or VIFF.

vertical so the Harrier can perform its unique reverse maneuvers. The Pegasus is able to accelerate the Harrier to over Mach 0.9, but not to supersonic speeds. Over the years many studies and several development programs have been undertaken to add supersonic performance to the Harrier's bag of tricks. Incidently, while the Harrier can take off vertically, due to its weight, this capability is seldom used except at air shows (Fig. 4-9). With fuel expended, the much lighter Harrier often lands vertically after the mission. Therefore, a replacement for the Harrier will probably have Short Takeoff/Vertical Landing (STO/VL) capability.

Fig. 4-9. Future V/STOL fighter will not only take off and land on very small flight decks, but probably will also have the capability to fly at supersonic speeds.

Thrust vectoring could find its way into many more aircraft. Indeed, the F-22 already has two-dimensional (2-D) capability (Fig. 4-10). Thrust vectoring is vital for enhanced agility or when flying at very high angles of attack where control surfaces lose much of their effectiveness. Thrust vectoring is also important for achieving short takeoff and landing capability, whether it be for a high-performance fighter or a business jet.

Initially, 2-D thrust vectoring used rectangular nozzles to obtain vectoring in the pitch plane mainly for STOL capability. Full three-dimensional (3-D), or pitch-yaw thrust vectoring was next (Fig. 4-11). Three-dimensional thrust vectoring could also be used in aircraft without vertical and horizontal tails, or at least in aircraft with much smaller rudders and horizontal trails. Beside producing signifi-

Fig. 4-10. The two-dimensional thrust vectoring exhaust nozzles are clearly seen on this F-22 Advanced Tactical Fighter.

cant drag, tail surfaces reflect radar beams and thus decrease stealthiness. An aircraft with vectored-thrust capability could land in much shorter distances by using a flare maneuver to bleed of speed in a manner similar to a helicopter. In both tailless and short-landing aircraft, thrust vectoring engine nozzles replace yaw and pitch control surfaces.

Thrust vectoring places new requirements on the flight control system because it must now integrate and control both the forces exerted by the engine and the traditional aerodynamic control surfaces. Complex algorithms are used to monitor the effect of pilot throttle, rudder, and stick inputs. Commands are sent to engine nozzle actuators as well as control surface actuators. However, the flight control system is totally transparent to the pilot, who flies the aircraft as if he is using control surfaces only. Of course now the aircraft has much more controllability and agility. The control system also has to prevent over-stressing the airframe. For instance, a turn input might call for 10-degree nozzle vectoring. If the airframe can only stand a maximum of 8 degrees, the nozzle responds by vectoring only 8 degrees.

Fig. 4-11. Three-dimensional "Axisymmetric Vectoring Exhaust Nozzle" (AVEN) provides both pitch and yaw vectoring. The AVEN is shown here during ground testing.

PROPELLERS ARE HERE TO STAY

In their neverending search for more fuel-efficient engines, engineers have shown a renewed interest in the propeller (Fig. 4-12). It is a well known fact that propellers on some large transports can achieve propulsive efficiencies of 85 to 87 percent. By comparison, the best of today's high bypass-ratio turbofan engines have efficiencies of 60 to 65 percent. Even the advanced turbofans of the future will never be able to approach the efficiency of the propeller. Interestingly, the propeller designed by the Wright brothers for their 1903 aircraft had a remarkable efficiency of 70 percent!

Unfortunately, the propeller-driven aircraft has traditionally been limited in its maximum speed to about Mach 0.6. As the speed of the aircraft approaches Mach 0.6, the airflow at the propeller tips reaches supersonic speeds and the efficiency drops drastically, not to mention there is a sizable increase in noise.

The renewed interest in propellers is centered around a new fuel-saving concept called the propfan, a design that appears to be quite practical now, mainly because of advances in materials technology. Indeed, the propfan may be the only radical change to be seen on commercial airliners and military transports of the 21st century.

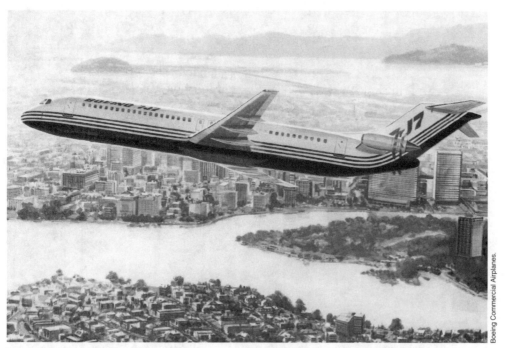

Fig. 4-12. Airliner concept that uses twin pusher propfans with contrarotating blades.

The appearance of the propfan differs substantially from propellers of the past. For starters, the propeller blades on the propfan are quite short. To provide the same amount of power many more blades will be used—8, 10, or even 12 blades may be common practice. The blades will be very thin with wide chords, which will increase the critical Mach number (the speed at which the blade reaches supersonic speeds). They will also be highly swept with a scimitar-like shape to further improve operation at high speeds. The blades will be made of advanced materials like composites and lightweight metallic alloys that offer both the high strength and low weight needed to keep centrifugal stresses down.

The small-diameter blades offer another advantage—a relatively small overall size, making for easier installation on the aircraft. This means the engines could be mounted at the rear of the aircraft and push, rather than pull, the aircraft. The pusher configuration is inherently more efficient than the more conventional tractor engine arrangement, and rear-mounted engines reduce the problem of cabin noise, produce less drag, and allow higher ground clearance. Also, placing the engines closer to the centerline of the aircraft greatly improves aircraft control characteristics if the aircraft has to fly with an engine out. The yawing and pitching is reduced under such conditions, and the need for excessively long fuselages and large tails to compensate for yawing is alleviated.

Engine manufacturers have considered designs with a single set of rotating blades as well as with two sets of smaller-diameter contrarotating blades (which can be installed close to the fuselage). Contrarotation also reduces gyroscopic effects and makes the engine even more efficient, perhaps about 10 percent more fuel efficient than a propfan with a single set of blades, if the engine is mounted at the rear of the aircraft.

Unless a conventional wing-mounted propeller is carefully designed, most of its swirl energy is lost to the airstream. The propfan's high RPMs coupled with its small blade diameter results in a great deal of swirl energy. By placing a second set of contrarotating blades behind the first set, much of the swirl energy from the first set can be converted to useful thrust by the second.

Some advanced designs incorporate a shroud around the propeller, effectively converting the propfan into a turbofan with a very high bypass ratio. The bypass ratio is a measure of how much air passes outside the core of the engine (i.e., the compressor, combustor, turbine, and nozzle section) compared to the volume that actually passes through the core. Turbofans with relatively high bypass ratios are already used extensively on military and commercial aircraft because of their good fuel economy at high speeds. The ducted propfan would extend this advantage and reduce acoustic problems.

The technology from propfan development will be useful on all future aircraft with propellers. And propellers will remain the most popular means of converting engine power to thrust, whether they are paired with turboprop engines, in the case of business and commuter aircraft, or internal-combustion engines for light and pleasure aircraft. These propellers, however, will take on new shapes and be made of new materials to make them more efficient, better performing, and quieter. Also sophisticated computer models developed for designing the new propfan engines will be adapted so that the best propellers can be designed for other types of engines.

THE ROTARY ENGINE

At one time, the rotary, or Wankel, engine was touted as the automotive power-plant of the future. Automakers invested vast amounts of money in its development. With early reliability problems and higher-than-expected fuel consumption, the engine lost its glamour. That is except for Japan's Mazda, who developed the rotary engine into a highly reliable and superb performing powerplant for its sports cars. Incidentally, the Mazda rotary engine has been used in some home-built aircraft. Between 1983 and 1991, NASA sponsored the Rotary Engine Technology Enablement Program, conducted by the Curtiss-Wright Corporation and John Deere and Company. The program was specifically aimed at light aircraft applications for an advanced rotary engine.

Most of the components in a rotary engine use rotary rather than reciprocating motion. There are no connecting rods, pistons, cams, valve trains, or other parts that move up and down. Thus, the rotary engine is much simpler with fewer components than reciprocating engines, resulting in greater reliability and easier maintenance. The rotary engine is also virtually vibration-free, runs smoother, and is quieter than internal-combustion engines.

Because the rotary engine produces three pulses for every turn of the rotor, high ratios of horsepower-to-engine displacement and horsepower-to-weight are the rule. Ratios of as high as 5-horsepower-per-cubic-inch were demonstrated in the NASA program. By comparison, a reciprocating engine that produces 1-horse-power-per-cubic-inch is considered very good. The design of a rotary engine results in a smaller frontal area with less drag (Fig. 4-13).

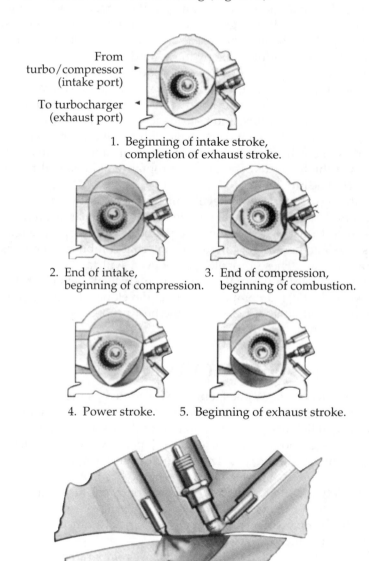

From
turbo/compressor
(intake port)

To turbocharger
(exhaust port)

1. Beginning of intake stroke,
   completion of exhaust stroke.

2. End of intake,
   beginning of compression.

3. End of compression,
   beginning of combustion.

4. Power stroke.

5. Beginning of exhaust stroke.

Fig. 4-13. The top figure shows how a rotary engine operates. The lower figures show the twin injectors used for stratified combustion in a rotary engine. Avco Lycoming.

While the rotary engine in itself shows promise, what really makes the engineers enthusiastic is stratified-charge combustion. In a stratified-charge engine, two nozzles, a pilot, and main injection nozzle are used to obtain the stratified charge. The pilot nozzle injects an overly rich fuel-air mixture into a portion of the rotary engine's combustion pockets cut into each of the rotor's three flanks. The pockets typically contain about 50 percent of the total working volume at top-dead center. The burning-rich mixture ignites a larger lean mixture that has been injected by the main injector. While several conventional automobile engines use a version of stratified combustion, the rotary engine is especially amenable to this technique.

Stratified combustion offers many advantages. First, it results in a more efficient engine, requiring less fuel and producing lower emissions. For aircraft makers, the biggest advantage is that the engine can be modified to use any one of many fuels, including Jet A, aviation-grade gasoline, automobile-grade gasoline, liquid petroleum gas (LPG), alcohol-based fuel, or even diesel.

Aircraft rotary engines will be turbocharged for peak performance and will most likely be liquid-cooled. With liquid cooling, cabin heating can be more efficient (as in a car) and safer, compared to the normal procedure of taking heat from the exhaust manifold on air-cooled engines. Routing liquid coolant through the leading edges of the airfoils could also result in an effective deicing system. Also liquid-cooled engines can make rapid descents from altitude, withstanding the sudden changes in air temperature.

While a single-rotor engine is possible, up to eight rotors can be stacked to obtain more power. Beyond that, the engine would become too long for aircraft use. For smaller commercial use, two stack-rotors with displacements of 200-cubic inches could produce 800 to 1000 horsepower.

DIESEL ENGINES

An interesting contender for light aircraft, and even business and commuter aircraft, is the diesel engine. At first glance, the idea of a diesel-powered aircraft might seem farfetched, but the concept has merit, and work on diesel aircraft engines was done by NASA and some engine manufacturers for several years. In fact, diesel-engine powered aircraft flew in the late 1920s and early 1930s.

Forget your vision of the familiar heavy, smokey diesel engine that powers the lumbering 18-wheeler down the highway. The aircraft diesel engine is quite different while operating on the same diesel cycle. A well-designed, turbocharged, two-cycle diesel engine made of lightweight materials and running at high RPMs can produce just as much power per cubic inch of displacement as the conventional gasoline engine. In addition, the aircraft diesel retains most of the characteristics that make it popular for ground transportation, including excellent fuel economy, rugged reliability, and fewer maintenance requirements. One of its primary attractions today is the fact that it uses fuel other than vanishing avgas. Diesel fuel is also safer. It is less of a fire and explosion hazard in a crash.

RECIPROCATING GASOLINE ENGINES

The conventional reciprocating engine, the most popular aircraft powerplant in terms of pure numbers, will undoubtedly remain the most popular light, general aviation engine well into the twenty-first century. A more cost-effective engine that meets all the demands of the light airplane (or automobile, for that matter) has yet to be found. Also, engine manufacturers have made huge investments in production facilities that would have to be drastically modified or even scrapped if a radically new engine became popular.

Piston engines for aviation have been far surpassed by automotive engines using dual overhead camshafts, multiple valves per cylinder, aluminum blocks and cylinder heads, and electronic management systems. For instance, aircraft engines still use antiquated control systems; systems using separate controls for manifold pressure, mixture, and engine RPM, and no control for spark advance at all. By contrast, automobile engines need to meet ever stricter government emission and fuel economy regulations while still delivering the performance customers demand. Automobiles now use very sophisticated electronic management systems for controlling ignition, metering fuel flow, and so forth.

Automotive control systems use sensors to precisely monitor engine operating parameters, such as temperature, ambient pressure, power output, etc. This information is then used to adjust ignition timing and fuel delivery for optimum power output while maintaining temperature, emission, and denotation within required limits. The bottom line is that today's automobile engines are smooth, powerful, clean-burning, and fuel-efficient over a very wide RPM range.

So why isn't all this advanced, but quite mature, technology found in piston-engined aircraft? Indeed, if it were, aircraft reciprocating engine's spark, mixture, RPM, and fuel flow could be controlled by a single lever, and fly-by-wire engine controls would be a reality. The aircraft engine would also deliver more power from a lighter weight package. One of the big reasons this technology is not found in piston-engined aircraft is the familiar product-liability problem.

Few aircraft manfacturers, or for that matter manufacturers of automotive engine control systems, are willing to install these new items that potentially lead to additional, and very expensive litigation. Advanced "smart" engines would also be much more expensive than current engine systems, partly because development and certification costs would have to be amortized over a very limited production rate compared to the high production rate of automotive engines. Also aircraft installations are more complex, especially because aircrafts require redundancy (or backup) by a duplicate electronic or mechanical system. Finally, the benefits of the "smart" engines are not as great in aircraft because aircraft engines operate over much narrower RPM and power bands, often running for hours at the same settings, whereas automotive engines encounter constantly changing conditions.

Turbochargers are used in both aircraft and automobiles. In earlier days, turbochargers were essentially added as an afterthought to conventionally designed piston engines. Newer automotive applications of turbochargers are more integrated with the engine design. Intercooling is one example of an improvement that can be made in turbocharged engines. When a turbocharger compresses intake air, heat is generated and the engine both loses efficiency and is subject to excessive temperatures. About one percent of efficiency is lost for every 10-degree rise in temperature. The intercooler is a heat exchanger that cools the intake air using cool air from the airstream. There are two passageways in the intercooler. One carries the intake air from the turbocharger, cooling it before passing it to the engine. The other passageway carries cool outside air (which reduces the temperature of the air in the first passageway) and sends it back into the airstream at a higher temperature.

AVIATION FUELS OF THE FUTURE

Fuels are an important part of the story of future propulsion. Fuel accounts for a major share of an airline's direct operating costs. Although aviation uses less than 10 percent of the petroleum consumed in transportation, this share is expected to grow to almost 15 percent by the year 2000. In the long run, fuel prices could again spiral upwards and fossil fuels are expected to eventually run out. Some experts believe sustainable sources of petroleum will not run out for at least 50 years, but by then virtually all of it will be coming from the politically volatile Persian Gulf region. The United States is projected to be one of the first countries to run out of petroleum. Therefore development of alternate fuels, and aircraft that can use them, is an important task for the future.

Of all potential energy sources, only three have been identified as viable alternatives to the jet fuels now used: synthetic Jet A (Synjet), Methane (CH_4), and Hydrogen (H_2).

Synjet is very similar to jet-grade kerosene except it is produced from coal, oil shale, or tar sands. Because its sources are fossil fuels, it is considered only an interim solution. There is also the problem of global availability, vital for worldwide airline operations. The United States is blessed with ample coal and oil shale deposits, unlike Western Europe and Japan, which would have to import fuel (or the materials to produce fuel) at higher costs. Coal and shale resources in the United States could last as long as 500 years. Concerns have also been raised over the environmental pollution caused by the strip mining of coal and the recovery of oil shale. Synjet's chief advantages are that it can be used in current engines (and with very little modification to fueling equipment) and can be produced at reasonable costs.

Methane is manufactured from coal, shale oil, natural gas, and biomass and, thus, is also subject to problems of global availability. It could be available in more countries if it was produced from biomass, but it is considered only a transitional fuel for the period between the use of Jet A/Synjet A and the development of hydrogen-fueled aircraft.

Methane would most likely be produced and transported to airports in gaseous form where it would be liquefied into a high-density cryogenic form for use on the aircraft. Substantial investments would be required for liquefaction plants and ground handling equipment, as well as retraining of personnel. And pollution would still be a worry because methane is a carbon-based fuel

Table 4-1 compares the alternative fuels discussed here and their specific characteristics.

Table 4-1. Advanced fuel availability in far future.

	Synjet limited	Methane limited	Hydrogen global
Renewable resource	No	No	Can be
Production investment	Moderate	High	Very high
Production cost	Reasonable	Reasonable	Very high (now)
Effect on aircraft design	Little	New Aircraft Needed	New Aircraft Needed
Effect on fuel handling	Little	Major	Major
Noise (compared to now)	Same	Same	Reduced
Pollution	Same	Same	Reduced
Safety	Same	Same	Improved

The most promising fuel for the twenty-first century is hydrogen. Its primary advantage is its global availability. It can be produced by numerous methods, many of which use renewable resources, and just about every country has some resources that could be turned into hydrogen. In several manufacturing methods, the primary ingredient is water. Large quantities of energy, however, are required to convert water to hydrogen, and convert hydrogen to the cryogenic form needed for aircraft fuel. One technique is electrolysis, in which any abundant source of electricity (e.g., hydroelectric, nuclear, solar, ocean wave, wind, or geothermal) could be used. Nations with coal, oil shale, or tar sand resources could make liquid hydrogen with fossil fuels. Other advanced techniques use biomass.

Hydrogen poses no pollution problems in use. The products of hydrogen combustion are just water vapor, nitrogen, and small amounts of nitrous oxides. In production, however, pollution is still an important consideration, especially if carbon-based fuels are used. The disadvantage of hydrogen is cost, both in capital investment and in processing. Today, no economical means to produce hydrogen in large quantities exists. As petroleum costs soar, the cost of producing hydrogen will become more competitive and the capital investment will be considered worthwhile. If fossil fuels are depleted, we will have no other choice than to turn to alternate fuels like hydrogen, regardless of the cost. Some experts already predict a "Hydrogen Economy" for the twenty-first century and much research is underway to find economical production methods.

One factor that might push ahead the use of hydrogen-fueled aircraft is the development of hypersonic aircraft, which will likely use hydrogen as fuel. Much of its technology and production facilities could be adapted for other military and commercial purposes. As aircrafts travel faster, much of the heat generated on the structure has to be cooled. Typically, this is done by passing relatively cool fuel, used as a heat sink, through the structure on its way to the engine. Jet fuel can be used and methane is even better, but hydrogen has the greatest potential for absorbing heat and, thus, it is the fuel of choice for flights greater than Mach 4 to Mach 6. Most likely, slush hydrogen, a mixture of solid and liquid hydrogen, would be used because more hydrogen within the same volume can be carried as slush than as liquid.

At first glance, safety might seem to be a major problem with hydrogen in light of the Hindenburg disaster. The Hindenburg, however, used hydrogen in gaseous form, contained within rubberized cotton cells. In the case of a hydrogen-fueled aircraft it would be in liquid form, contained in welded metal tanks. Granted, hydrogen is a high-energy fuel and must be handled accordingly, but safety experts give it high marks and in case of a crash, passengers and surroundings would be subjected to less hazard than with other types of aircraft fuels. Interestingly, only about a dozen people were killed in the Hindenburg accident, yet it doomed the airship as a means of air transport. If the same logic had been applied to airline travel, people would not be flying today.

5

Avionics

THE TECHNOLOGY that will have the greatest impact on aircraft of the twenty-first century will be avionics, the electronics used in aviation. Avionics started with simple radio voice-communication but took large leaps forward with military aircraft, missiles, and the space age. In no other area have the advances in military and space technology been more rapid and widespread than in electronics. Every sector of aviation will be affected by future advances in avionics. Advanced avionics can even be incorporated into old aircraft, often with a simple installation (Fig. 5-1).

Fig. 5-1. The venerable DC-3 equipped with the latest in avionics, bringing it from the 1930s into the 1990s, maybe even into the twenty-first century.

New uses for electronics are almost unlimited. While most commercial aircraft flying in the twenty-first century will look much like the airliners in use today, and might, in fact, use today's airframes, even a casual look into the cockpit will reveal some major changes.

For the military, advances in electronics will not only mean new aircraft with new capabilities, but also existing aircraft with updated capabilities. In the future, electronics alone can change the characteristics of an aircraft. For example, Swedan's Saab-Scania has developed its JAS-39 Gripen as a multipurpose fighter designed for fighter/interception, ground attack, and reconnaissance missions (Fig. 5-2). To change roles, the pilot needs only to select the appropriate computer software with the press of a button in the cockpit. Enhancements to future aircraft would be ensured by adding new computer hardware and software. This growth capability will take on even greater importance as aircraft developments and life-cycle costs become more expensive and defense budgets become more austere.

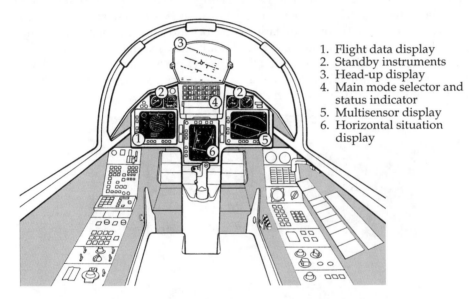

1. Flight data display
2. Standby instruments
3. Head-up display
4. Main mode selector and status indicator
5. Multisensor display
6. Horizontal situation display

Fig. 5-2. The displays and controls in the Saab JAS 39 are ergonometrically laid out to reduce pilot workload. Saab-Scania.

Avionics' impact will probably be felt the most in the general aviation market. For greater safety, small aircraft will be required to carry rather sophisticated electronics. Fortunately, the price of new electronics has a tendency to go down as the capabilities of the electronics go up (the home computer is the best example of this). Avionics that are now found only in military and large commercial aircraft will become available to the pleasure flyer at reasonable costs. One example of this is the Global Positioning System (GPS) equipment that has come down in price so that it can be afforded by most private pilots.

Electronics will also take on tasks now done by mechanical equipment, providing weight and cost savings, as well as improving reliability and maintainability. This shift to electronics has already been seen with the widespread use of fly-by-wire flight control systems that replace mechanical actuators, cabling, and hydraulic lines. In the future, electronics could eliminate or reduce the use of hydraulic, pneumatic, mechanical, and accessory gearbox systems. Electronic substitutes include ultrareliable electrical motor drives and controls that would use miniaturized, high-power, solid-state switches based on emerging metal oxide semiconductor thyristor (MCT) technology. Besides allowing substantial weight reduction, the electronic replacements would be less expensive and more resistant to combat damage.

Not all mechanical systems will be replaced by electronics, and the ones remaining will be designed for more efficiency and less weight. For example, a new electric starter/generator would be internally mounted directly on the centershaft of a turbine engine, doing away with separate gearboxes, hydraulic pumps, constant speed drives, and starters.

THE BUILDING BLOCKS OF AVIONICS

The key technologies that will lead avionics development can be categorized into three broad areas: computers, optics, and artificial intelligence.

Computers

Future aircraft will require tremendous increases in computer power. One advanced computer technology is already being installed in the latest aircraft, as well as being used to update earlier craft. This technology is the very-high-speed integrated circuit (VHSIC), the result of a decade-long, billion-dollar investment made by the Department of Defense.

By the 1970s, the electronics industry had turned its focus from the relatively small military market to the huge consumer market. Not only were the profits greater, but the demands were less stringent. Commercial applications usually did not require the expensive testing and quality control required in such applications as satellites, missiles, and high-performance aircraft. The situation reached the point where the military had to use computer technology developed for the civilian market, but they couldn't use it for quite a while after it was available to commercial users because of the modifications and extra testing needed for military use. The U.S. military, recognizing American superiority in computer technology, and realizing the military was falling behind in developing future systems, established the VHSIC development program. What the military required was vast improvements in computational speed for radar, weapons targeting, electronic warfare systems, and so forth. Because VHSICs are designed specifically for military applications, they incorporate features for survivability on the modern battlefield, including immunity to radiation and electromagnetic pulses. Eventually, VHSIC technology will permeate everything from airplanes and missiles to submarines and helicopters, as well as satellites and ground-based communication systems. The applications are virtually unlimited.

Another important future step in microchip technology might be multichip modules (MCM). Conventional printed circuit boards use individual integrated circuits packaged separately and connected with copper wires, resulting in a rather inefficient use of the circuit board area. Typically only about 5 percent of the total area is devoted to the actual circuit. In a MCM, unpackaged integrated circuit (IC) are mounted on the substrate and connected by many very fine wires to allow as many chips as possible on one module. The benefits of MCM, with its increased IC densities, include vast increases in speed and reliability along with reduced size and weight. Some experts predict that by the year 2000, more than a billion gates could be integrated on a single MCM, meaning a complete and complex computer system could be placed on a single 6-inch square by 1-millimeter thick MCM.

Today, integrated circuits are manufactured using optical lithography with an optical electron or ion used to etch the pattern on a silicon wafer. This technique is physically limited when it comes to the size of features that can be etched, as well as the ability to produce patterns simultaneously over large areas. X-ray lithography, which operates at shorter wavelengths, can produce semiconductors with a quarter micron and sub-quarter micron feature sizes. X-ray lithography also allows more devices to be placed on a given wafer size. Denser spacing leads to increases in the operating speed of the circuit and even brings the price down.

While much superconductor research is aimed at superconducting magnets for use in very efficient generators and motors, scientists are also working on high-temperature superconductors (HTS) for computer applications. High-temperature here means a class of materials like tallium-barium-calcium-copper-oxygen that superconduct at temperatures of below 105 degrees Kelvin or yttrium-barium-copper-oxygen that is superconducting up to 89 degrees Kelvin. While not really warm by conventional standards, 32 degrees Fahrenheit equates to 273 degrees Kelvin, they represent a significant improvement over early superconductors that only worked near absolute zero. The HTS can be cooled using simple liquid nitrogen coolers since the temperature of liquid nitrogen is 77 degrees Fahrenheit. Superconductivity results in a lack of electrical resistance, so superconductors dissipate very little power and generate essentially no heat. HTS could be used to fabricate electronic circuits that are significantly faster than conventional circuits made with copper or gold.

Future computer chips could also use gallium arsenide (GaAs) in place of silicon. GaAs chips have long tantalized electronics designers with its promise of much greater speed, greater tolerance to large temperature changes, and the ability to absorb more radiation without any deterioration. But GaAs also has frustrated them with its fragility and production difficulties.

The key to super-speed computers is how the chips are integrated. Time-saving, efficient computer architectures and large memory capacities allow high-speed chips to work at their maximum speed. Terms like *scalar*, *vector*, and *parallel processing* are used to describe computing techniques. The first electronic computers were scalar machines in that they worked on one number at a time. In the 1970s, Cray Research introduced the first computers that worked on a number of multielement vectors at once, which is the way most problems in science and engineering are formulated. The Cray 1 provided a huge increase in problem solving

The futuristic Beech Starship I in flight.

The Seawind is a futuristic, light seaplane design.

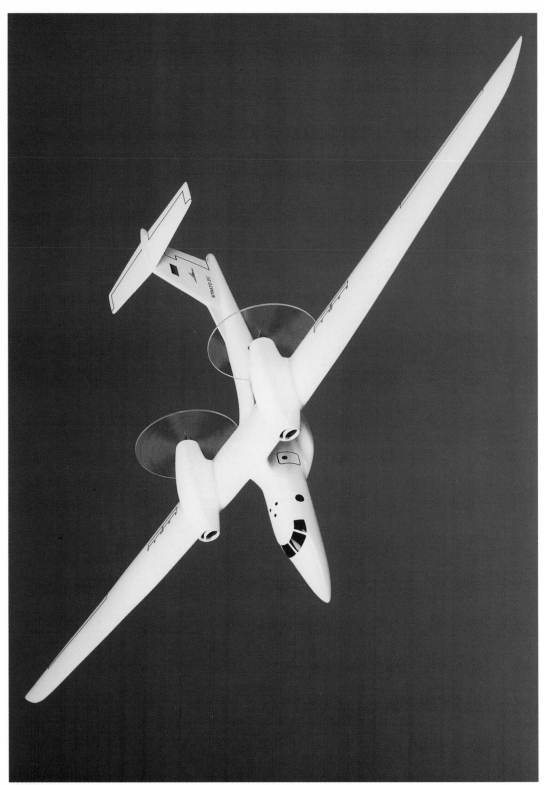

The two-man Strato 2-C could stay aloft at altitudes of about 85,000 feet for days studying the earth's atmosphere. Power would be supplied by twin turbocharged engines. Deutsche Forschungsanstalt für Luft and Rahrfahrt

The F-22 Advanced Tactical Fighter will be the U.S. Air Force's first-line air superiority fighter well into the twenty-first century. Lockheed Aeronautical Systems Co.

The X-30 National Aerospace Plane could eventually demonstrate single-stage-to-orbit flight using air-breathing engines as its primary means of propulsion.

NASA

Artist's concept of an advanced military helicopter that would use circulation control to completely eliminate the need for a tail rotor. Hughes Helicopters

The F-117A Stealth Fighter proved itself during the Persian Gulf War. USAF

The F-22's stealth characteristics can be clearly seen in this front view.

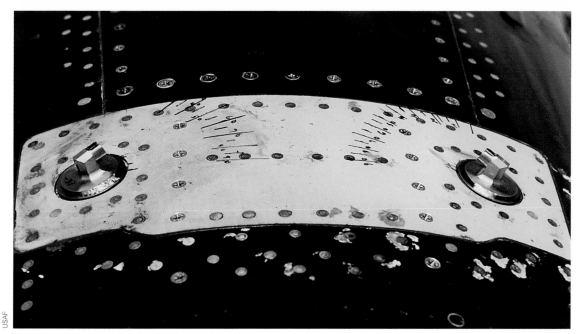

Close-up of VFC nozzles on the X-29.

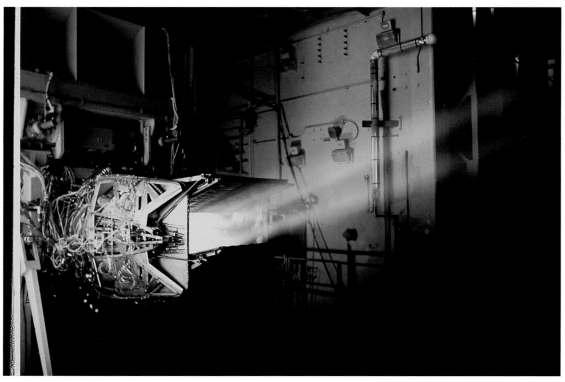

Two-dimensional thrust vectoring nozzle in tests. Two-dimensional nozzles provide pitch control.

The C-17 can carry large pieces of military equipment to the battlefield.

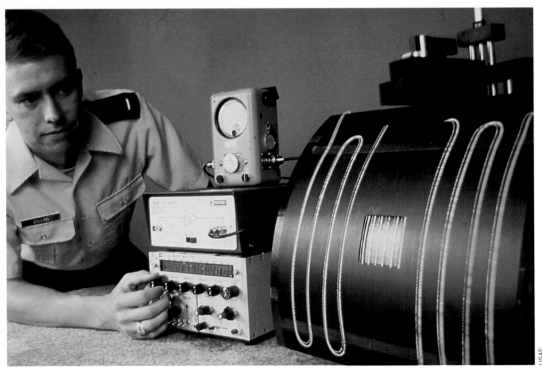

Fiber-opitc material is embedded in composite material to create a smart skin. Smart skins can be used as antennas, sensors, or to monitor the health of the structure.

speed when it was introduced in 1975 with a sustained capability of about 150 Mflops (millions of floating point-operations per second). The Cray 2 of the mid-1980s had a peak speed of 2000 Mflops and a sustained speed of 250 Mflops. To show how computation speeds jump by orders of magnitude, computational speeds are now expressed in terms of Gflops or gigaflops, where one Gflop is equivalent to 1000 Mflops. Cray's Triton project is aimed at a peak performance of 60 Gflops by 1995–1996.

Parallel processing is becoming the latest high-power tool for supercomputing. With parallel processing, the problem is split so parts of it are tackled simultaneously by many processors. This requires some sophisticated componentry to make the various processors work in unison without wasting time when exchanging data or programs. Generally, parallel processing is discussed in terms of massively-parallel computers because they use several hundred or even thousands of processors, which incidentally are usually adapted from mass-market personal computers or workstations in order to keep costs down. To express the capability of the machines of the future, teraflops, or 1000 gigaflops, are used to describe some parallel processing machines. One teraflop sustained speeds are expected by around 1997.

As important as the computers themselves are, the software that turns inanimate electronics into problem-solving brains is of equal importance. Often as much or more is invested in computer programming for a new system as is invested in the hardware. For aerospace applications, much of the cost of developing a new program is in checking it out under virtually all possible operational situations. When safety or mission accomplishment is involved, the possibility of a "bug" in the computer software cannot be tolerated.

Future aircraft will use not one, but several computers. Not only must the computers communicate with one another, they must be able to interact with computers aboard other aircraft, space satellites (that might be providing communications and navigation information), and computers on the ground. To add to the complexity, the computers must share a common language. The computer community, therefore, has expressed great interest in adopting a common language for future computer systems. That language is Ada.

Optics

Fiberoptics is one of the most promising of the newly emerging technologies. Amazingly large amounts of information can be transmitted at the speed of light along fiberoptic strands no larger in diameter than a human hair. The information is in the form of pulses of light called *photons*, which is sent in a manner akin to a flashlight sending Morse code down a long piece of pipe. With fiberoptics, the light source is the laser, and many messages can be transmitted simultaneously by using different frequencies of light. In theory, it would be possible to transmit all of the world's telephone conversations on a single optical fiber. And the speed of photonics, as this new technology is called, is such that the entire contents of the Library of Congress could be transmitted in a few seconds.

Fiberoptic cables are also compact, energy efficient, and immune to electromagnetic radiation interference. The latter has important military implications in that, unlike fly-by-wire flight control systems, which are quite susceptible to intentional or unintentional interference, fly-by-light systems would not be affected.

Fiberoptic gyroscopes, or FOGs, are beginning to replace the venerable mechanical gyros. Advantages of FOGs include a compact size, increased durability because moving parts are eliminated, reduced power requirements, and a significantly lighter weight. In one inertial measurement unit (IMU), a FOG reduced weight from 25 pounds when a mechanical gyro was used to a mere 15 ounces. The FOG is based on the Sagnac effect discovered in 1913 by the French scientist after which it is named.

Laser light is split so it travels in opposite directions, for instance, in a circular path. When the gyro rotates, a very small increase in path length in one direction results and an equally small decrease in the other direction. When the two beams meet, this relative difference in path length can be detected as a phase shift which, when converted to outputs, can be used in the same way as if they came from a conventional mechanical gyroscope.

Later applications of photonics will probably include optical computers that could operate thousands of times faster and more efficiently than today's microchip-based machines. Optical scanners that can "interpret" photo imagery would have important applications in reconnaissance. Some of the experts even believe that photonics are as technologically revolutionary as the transistor was in the 1950s.

Artificial intelligence

Another technology revolutionizing the aerospace and computer communities is artificial intelligence (AI). The term *artificial intelligence* was coined in 1956, and the idea itself has been around since 1947. Artificial intelligence is that branch of computer science that attempts to use computers to solve problems in a manner that simulates the human reasoning process, in other words, to "think." Expert systems, neural networks, machine intelligence, natural language, and machine reasoning are forms of AI that have been investigated and used with varying degrees of success.

In an expert system, computer databases in a memory bank contain as much knowledge on a subject as can be compiled. In other words, the system is an "expert" on a particular subject and can be consulted by less knowledgeable experts, or even nonexperts, to do a job or make an assessment of a situation. For example, the expert database in a cockpit system might include such information as engineering and system data, weather and navigation, threat assessments, and the seasoned judgment of experienced pilots. The actual decision-making is done using a "rule-of-thumb" approach based on expert experience. Decisions are made differently from the conventional right-or-wrong method of a binary computer. The expert system need not offer an absolute answer but can indicate the preferable alternative—the one with the highest probability of being correct—thus imitating the decision-making process of the human brain.

Expert systems offer many ways to help commercial pilots overcome their ever-increasing workload. The "Diverter," developed by Lockheed Aeronautical Systems Company for NASA, shows one application of AI in the airliner cockpit. The system helps an airline pilot solve an in-flight problem such as a sick passenger requiring a diversion from the normal flight plan with minimum impact on his job of piloting the aircraft (Fig. 5-3). Inputting a few commands into a computer next to the pilot, the Diverter quickly recommends the nearest open landing site that can accommodate the special needs, displays a color map to the new destination, and gives the pilot the reasons this destination was chosen. Diverter uses a processor with software using AI concepts and programming techniques to weigh all alternatives in a matter of seconds, make a decision, and produce the rationale. The onboard database includes the airline's route structure, schedules, and carrier-specific policies, plus information on airfield facilities, runway length, proximity to hospitals, availability of emergency equipment, terrain, etc. A data communications link with surface stations provides real-time information on changing weather and airfield conditions.

Lockheed Aeronautical Systems Company.

Fig. 5-3. The Diverter program uses artificial intelligence concepts to provide pilots with critical information should the aircraft have to be diverted to an alternate landing site because of an in-flight emergency.

In an increasingly complicated, high-risk combat environment, military pilots need all the help they can get, so the military is investing vast resources in a form of AI, the Pilot's Associate (PA). PA serves as an additional aircraft "crew member," handling much of the routine workload. For example, when a fighter detects and engages the enemy, the PA would recommend the most effective attack profile for the aircraft and its weapons. It could preset the weapons while positioning the aircraft for the best attack advantage and, if commanded, could even execute the attack. Throughout the engagement, the PA could increase survivability by recommending and, if desired, even executing appropriate defensive counter measures and evasive maneuvers. However, as in all AI applications, the pilot would still have complete override capability (Fig. 5-4).

Fig. 5-4. The Pilot's Associate assists pilots of fighter aircraft in processing the immense amount of data from both inside and outside the aircraft.

Neural networks are a potentially powerful means of making decisions about data using a technique inspired by the way neurons work in the human brain. In either a computer or the brain, large amounts of information is handled by breaking it into tiny, manageable pieces. In the computer, data is handled by gates that fluctuate between "on" and "off." In the brain, the "on-off" is paralleled by neurons firing on or firing off. However, even though the neural network is similar to the brain, the network does "think" like the human brain.

In contrast to a standard computer program that uses detailed rules and instructions to arrive at solutions, neural networks are not programmed in the traditional sense. Neural computing does not follow a strict and rigid set of lengthy instructions. Neural networks are trained to make decisions by learning by example and self-organizing. In the process, neural networks discover the rules for themselves through training. Examples of data are repeatedly introduced to the network, which learns from its errors to the point that it trains itself to consistently produce correct decisions. When the neural network guesses wrong, it adjusts its internal connections until it gets the right answer, and then it "remembers" this.

Neural network technology is especially effective with real-world data that is not well behaved. It is a quicker, easier way to do something that in the past, people have struggled to do because of the complexity of the data and the slowness with which previous mathematical techniques offered a solution. Instead of being massive number crunchers, the networks are especially adept at pattern recognition, such as for weapons targeting and maneuvering, and for problems with real-world ambiquity.

AI-based fault-diagnosis systems offer some interesting possibilities. Besides identifying a problem, diagnosing the fault, and determining its effect on the flight, an AI system could actually reconfigure the malfunctioning components, working around the difficulty so that the flight could be safely continued.

Much technology needs to be developed, however, before AI is a routine reality in the fighter or commercial airliner cockpit. Foremost is the tremendous amount of computer power required to be packaged in units that will fit in the cockpit. Then there are the very sophisticated software packages that must assimilate all the data required for an expert system or neural network to present intelligent choices to the pilot.

COCKPITS OF THE FUTURE

While some new avionics will have to wait for entirely new airplane designs, others will be incorporated into aircraft flying today as soon as they become technically and economically feasible.

Airliner cockpits

In the twenty-first century airliner cockpit, sidestick controllers, like those found in fighters, could replace the familiar control wheel and yoke. With fly-by-wire or fly-by-light control systems, the mechanical advantage provided by conventional control yokes is no longer needed. Elimination of these controls greatly improves the visibility of the control panel, frees up space for other controls and instruments, and enhances control of the aircraft with minimum physical exertion.

Although not related to avionics, another big improvement will be to windshields, improving the view of the outside world, and thus enhancing collision avoidance. For example, future windshields could be one piece and made of strong plastics that could better withstand bird strikes.

The biggest changes, however, will be found on the instrument panel. Advanced panels will be designed ergonomically to greatly reduce pilot workload. Until recently, flight decks grew in a rather hodge-podge manner through the replacement of outdated components and instruments with more modern ones as they become available. This resulted in a conglomeration of knobs, switches, and electromechanical displays contributing to excessive crew workloads and, possibly, misinterpreted information. Head-up displays (HUD) are finding their way into airliner cockpits. HUD can provide an extra margin of safety during takeoffs and landings, for instance, when it is important that the pilot keep eyes on both the view outside and his instruments.

The Electronic Flight Instrumentation System (EFIS), now appearing on the commercial avionics market, will go a long way in correcting this problem. With the EFIS, all flight and performance information can be displayed on a few cathode-ray tubes or flat-panel, liquid-crystal displays. The key part of the EFIS is the computer-generated graphic display. Multicolor, high-resolution displays are available. Multiple windows that can display a variety of information simultaneously (eliminating the need to scroll through computer menus) are already available. The almost limitless potential for computer graphics includes computer-generated maps, route and destination charts, and real-time multicolor weather displays.

The instruments of the twenty-first century airliner will provide great strides in safety and economy. Avoiding midair near-misses is of great importance, and systems, such as the Traffic Alert and Collision Avoidance System (TCAS) are now being developed to help the pilot. TCAS requires transponders to be installed in cooperating aircraft. TCAS then interrogates the transponders in all nearby aircraft. From the responses, TCAS can compute their range, altitude, and bearing. If one of these aircraft is in the airliner's airspace, the pilot is warned both visually and aurally. The pilot is also given instructions on the least-disruptive avoidance maneuver to assure adequate separation. TCAS can also continuously display traffic data on the weather radar screen to aid the pilot in tracking his airborne neighbors.

The Wind Shear Detection/Alert System (WSDA) is another important safety device. When WSDA detects an unsteady air mass, it gives the crew an amber wind-shear warning. The system can detect wind shears in sufficient time for the crew to take corrective action.

After safety, maximum profits are next in importance to the airliner. Thus, systems like the Flight Management Computer System (FMCS) will become an integral part of the flight deck. FMCS uses a computer (integrated with the autopilot and engine controls) to obtain optimum fuel economy and overall aircraft operating efficiency.

Military cockpits

When it comes to the cockpits of military fighters, bombers, transports, helicopters, and V/STOL, as the saying goes, "you ain't seen nothin' yet."

The Head Up Display (HUD) is already a way of life in high-performance military fighters. Likewise, helmet-mounted displays (HMD) are routinely used in military helicopters. Virtual Cockpit takes the idea of the HUD and HMD one step further. In combat, a pilot needs a 360-degree view to keep track of both targets and threats. At the same time, he needs instantaneous information on critical aircraft parameters. With a Virtual Cockpit display, the pilot has a panorama of information from the aircraft's sensors and avionics, organized both in time and in three-dimensional virtual space. This display might be produced by miniature cathode-ray tubes mounted on the pilot's helmet and projected through the visor optics into the pilot's field of view. Head-tracking technology could determine where the pilot is looking at a particular instant so that the data could be pointed in the direction he wishes to view. A critical technology item in these types of situations will be a helmet that can handle all the optics and electronics and still be light and comfortable enough to wear for long periods.

HUDs and Virtual Cockpit displays could present more than just information about aircraft systems. Using sensors like forward-looking infrared cameras, a pilot could "see" through rain, fog, snow, and darkness by viewing an electronically produced image of the scene outside. When he breaks out of the clouds there would be a harmonious blending of the electronic image with the real picture. You can probably see why this idea has been dubbed the "Magic Window." Such a system would not only be great for military pilots who must fly in all types of weather, but also for commercial carriers.

An even more advanced concept is called the *God's-eye view*. Here, the pilot would be presented with a big-picture view of his combat situation. Projected on the HUD or HMD, the view would be the one he would see if he were located at a point outside the airplane, such as above and behind his aircraft. He would get a better view of threats, friendly aircraft, targets, and ground systems, and his relation to them.

Because of the intense workload in future aircraft cockpits, voice control will be a great help. Pilots will verbally command cockpit functions, such as dialing radios and selecting weapons to be fired. The pilot will only have to speak the appropriate command and watch the appropriate display to determine that the verbal command was executed. Voice control could even be coupled to the Virtual Cockpit so that the pilot would only have to look at a switch or display and utter a word or two to make things happen.

Voice recognition systems must correctly identify words as spoken not only by different people, but under varying conditions. For example, a pilot would probably have a significantly different speech pattern under stressful situations, such as in the heat of combat or during an inflight emergency. Voice control systems could use sophisticated schemes to recognize disparate voices, or could be a bit simpler by only recognizing the voice of the individual who trained it. To make voice commands valuable, recognition technology has to be developed to the point where it can handle normal conversational tones and speaking speeds, instead of requiring slowly enunciated words. Also, vocabulary has to be expanded beyond the limited ones used in initial demonstration projects.

General aviation cockpits

General aviation aircraft often fly in and out of small airports that might not even have a control tower. These airports could have completely automated traffic-control systems that could provide pilots with vital information on weather conditions, airport characteristics, and air traffic without a single human involved.

Radar would track air traffic, and weather sensors would gather climatic data. This information would be fed into a minicomputer that would convert it to voice-synthesized reports that would be transmitted on an assigned radio frequency. Or the information could be displayed on a CRT or LCD on the instrument panel. Several times a minute, a pilot could get the tail number and location of all aircraft within a 3- to 5-mile radius of the airport. Every few minutes, weather information would be broadcast along with updated airport advisories. Such a system could make flying into and out of small airfields much safer, and without the high cost of labor.

SENSORS

Sensors are the "eyes and ears" of an aircraft. Sensors can be as simple as a pitot tube, which measures airspeed, or as complicated as a forward-looking infrared (FLIR) camera, which can "see" in total darkness. A multitude of sensors are needed to generate the information presented to pilots on CRTs and LCD displays.

Many of the advances in sensor technology will center around techniques that will allow aircraft to fly and fight in less-than-ideal conditions. For example, the FLIR cameras, unlike radar, are passive sensors that do not emit radiation that can be detected by the enemy. And while radar will still be a primary form of sensor, new radar will operate at millimeter wavelengths, allowing it to see through the fog, rain, and smoke that clutter the view of normal radar.

The skin and structure of future aircraft can serve as integral parts of its avionics. Sensors and antennas can be embedded in an aircraft's Kevlar and fiberglass skin and structure. For example, VHSIC microcircuits can be etched on the skin surface by using photographic techniques.

By combining microcircuits, sensors, and antennas embedded in the skin with other advanced technologies such as fiberoptics and miniprocessors, entire systems like radar, electronic countermeasures, and radar warning receivers can be moved out of the interior of the aircraft and into the structure and skin. The same apertures can be used for several functions from the same unit, such as offensive systems like radar and defensive systems like a radar warning receiver. Skin-embedded sensors would be set up in array configuration, so that sensors with similar functions will be located all around the aircraft. Array configurations allow for "graceful degradation," meaning failure or damage to one part of the system would not cause sudden failure of the entire system. Instead, failure would occur over a long period of time.

These "smart" skins and structures are natural for stealth aircraft because they eliminate protruding electronic pods and domes that produce telltale radar signatures. Smart skins can also reduce the weight and size of avionics systems, thereby increasing the number of systems an aircraft can carry and decreasing the aircraft's size. As electronic components get smaller and smaller, dissipating the heat they generate becomes an ever increasing problem. Putting tiny electronics into the skin of the aircraft lets them be cooled by the external airflow.

"Smart" skins and structures embedded with a maze of distributed sensors can also monitor the structural health of an aircraft while in flight. The pilot can be warned of a problem and can take appropriate action. For example, if a fighter receives combat damage, the pilot would now know he has only a three-G capable aircraft and not a full nine-G fighter.

HELP FROM SPACE

Military missions are already heavily dependent upon information gained from orbiting spacecraft. However, now most of the information must pass through ground terminals before it reaches the ultimate user in the cockpit of a fighter, bomber, or transport. This will change in the future. Navigation, intelligence, weather, and other data will be transmitted directly from satellite to aircraft, eliminating critical time delays. As equipment costs fall, this capability will be available to commercial users and then to the general aviation population.

The Global Positioning System (GPS) has brought a major improvement to navigation in the air as well as on water and land. GPS consists of a 24-satellite constellation in geostationary orbit. Since GPS is passive, it can be used simultaneously by an unlimited number of users, and their locations are not revealed when they use the system. It can be used under all weather conditions. The user's onboard equipment automatically computes the user's precise location.

Many airlines are already installing airborne satellite communications systems that can "talk" to ground stations via satellites in geosynchronous orbit. As many as 3000 air transports could be equipped with satellite communication systems by the turn of the century. Satellite communications also allow airline passengers to communicate while in flight via telephone, fax machines, or computer modems.

Much more information of value to pilots could eventually be obtained directly from space. Up-to-the-minute weather information could be transmitted from meteorological satellites directly to the cockpit. Someday, air traffic controllers will depend on radar and other sensors aboard space satellites to keep track of air traffic, and will use communications satellites to transmit messages, both by voice and picture, directly to pilots. Eventually the entire system could be automated with computers, computer-generated graphics and voices, and voice recognition systems, replacing human operators.

PUTTING IT ALL TOGETHER

In the past, aircraft have pretty much relied on separate equipment for individual functions, such as radar, communications, navigation, etc. Avionics seemed to be stuffed into airframes almost as an afterthought. This led to duplication of parts and excessive weight, cost, complexity, and maintenance problems.

Now avionics are being designed as an integrated package providing timely and easy-to-understand information to the crew on its CRTs, LCDs, and HUDs. Common components are used wherever possible, which not only reduces the required inventory of spare parts, but also leads to lower costs due to larger production runs. Black boxes that weighed up to 60 pounds can eventually be replaced by video-cassette-sized modules that weigh about a pound. By using self-diagnosing avionics, the crew can be alerted when a particular module is malfunctioning and might even be able to replace it with a spare diskette stored aboard. Maintenance people will have an easy job identifying faulty avionics and can make repairs by slipping in a new module. Fewer connectors and cables will be needed to tie systems together, reducing weight and increasing reliability (Fig. 5-5).

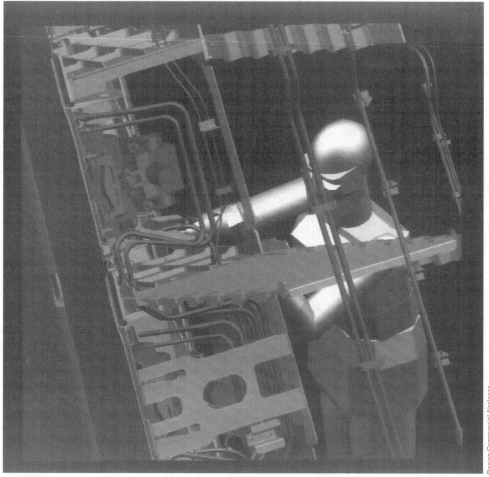

Fig. 5-5. Boeing Engineers used sophisticated computer programs to design the 777's structures and simulate the assembly of an individual part with neighboring parts and systems. Then "Catia Man," a computerized model of a human mechanic, is inserted into the electronic mock-up to make sure there is adequate space to service the airplane.

Self-healing avionics could be an important step towards increased ability to complete missions. By using common components, it will be possible to automatically reconfigure the avionics so that the electronics of another system can be substituted for the broken component to keep things going.

SIMULATORS

Because of the expense of flying aircraft for training and practice, simulators that closely duplicate actual flight without leaving the ground will be used increasingly in the future. Also, many new design concepts will be tested on ground-based or airborne simulators.

Flight simulators

Flight simulators have come a long way from the Link Trainer of World War II. Today, fledgling pilots can gain much of their training in very sophisticated simulators. Experienced pilots can maintain their proficiency without burning expensive fuel, and can even be upgraded to fly new aircraft by doing most of their training in a simulator.

Combat missions can be simulated so that expensive ammunition and missiles need not be wasted in training, and the aircrew can be given immediate feedback on their "gunnery" scores. By teaming up pairs of simulators, it is possible to practice air-to-air combat (Fig. 5-6).

Fig. 5-6. This Weapon Systems Simulator allows pilots to fly the latest fighters in realistic flight environments. Computer-generated images are projected inside the domes and the pilots do their "flying" from cockpit mock-ups that duplicate the ones in real aircraft.

Simulators allow crews to practice emergency procedures that are too dangerous to attempt in a real aircraft. You would not want to practice flying a supersonic fighter with a broken rudder, or landing a commercial airliner with its wheels up. But these are easily done with a simulator, and if you crash, you merely press the reset button and try again until you get it right. Also, by recording the simulation, the pilot can see a replay of mistakes that might have been made (Fig. 5-7).

Fig. 5-7. This photo shows how the images are developed for a flight simulator. In the first computer screen (left), the actual model is digitized. In the second screen (center), it is turned into a three-dimensional, geometric model. In the third screen (right), it is enhanced for inclusion in the dome's program.

Simulators are becoming more sophisticated to give pilots the most realistic simulation possible. With simulators that move, aircrews can experience all the sensations of actual flight, such as buffeting in turbulence, maneuvering in combat, and rough landings. Using special "G" suits and seat cushions that inflate and deflate rapidly, large acceleration forces can be simulated. Tape recordings and loud speakers duplicate normal sounds heard in the cockpit, and computers can synchronize things so every sensation happens at precisely the same instant for a realistic simulation.

Virtual reality could make simulation training much more realistic. Virtual reality can be best explained as an "experience" rather than just a sound or image. Rather than looking through the computer screen window at the data, you are on the other side of that window inside a world created entirely in a computer. Current state of the art in virtual reality allows an individual to have a multisensory experience that includes three-dimensional sight and sound, plus touch, feedback, or forced resistance and motion. Input to the computer and sensory feedback can be accomplished with data gloves, joysticks, helmet-mounted displays, goggles, earphones and body suits connected to graphic workstations.

Virtual reality provides very realistic training for pilots via simulators. It is also promising in developing maintenance and repair skills on equipment that does not yet exist. Technicians could perform repairs on a virtual aircraft, going through the exact steps they would use in repairing the actual aircraft. They could practice repairs years before the actual aircraft is built, allowing potential problems to be identified and corrected while the aircraft is still in design.

6

Materials and manufacturing

NO MATTER how good an aerodynamic design or a propulsion concept might be, it is not of much use if materials are not available to transform it from a laboratory experiment to a practical reality. The history of aviation is filled with great ideas that had to wait until the right materials came along to make them work. The forward-swept wing and the scramjet are two examples that immediately come to mind.

The vocabulary of the aerospace world already includes such terms as *composites, superalloys, ceramics, carbon-carbon, intermetallics,* and *metal-matrix* composites. Many of these materials are already in use. However, advanced materials are still in their infancy. The realization of their full potential is in the future.

Manufacture of new airframes, engines, and black boxes must be done economically and in sufficient quantities. Manufacturing has entered a major revolution with automated and robotic factories that build items designed on a computer screen, not on a drawing board. Aircraft are being developed with computer-aided design (CAD) and computer-aided engineering (CAE) techniques, then built in highly automated aircraft and engine plants using computer-aided manufacturing (CAM) (Fig. 6-1).

METALS AND CERAMICS

For decades, aluminum has been the mainstay of the aerospace industry, and the suppliers of aluminum products are trying to keep it that way. Future aircraft, however, must be lighter to achieve improved fuel economy. If the aircraft travels at very high speeds, aerodynamic heating can produce temperatures too high for conventional aluminum.

Aluminum alloys

To see how new forms of aluminum will impact the aerospace industry in the twenty-first century, just measure how much aluminum is used in a typical commercial airliner. Today, more than 70 percent of the aircraft is made of "ordinary"

Fig. 6-1. This huge Automated Spar Assembly Tool (ASAT) is used to produce the wing spars for the Boeing 777.

aluminum alloys. By the year 2000, new lightweight and higher-strength advanced aluminum alloys could comprise more than 50 percent of the airliner.

One of the most promising aluminum alloys is aluminum-lithium. Aluminum-lithium is both about 10 percent lighter and stiffer than plain aluminum. When aluminum-lithium is substituted in future designs, overall weights could be reduced by as much as 15 percent. Besides being lighter, aluminum-lithium has superior fatigue resistance, meaning components made of aluminum-lithium could last two to three times longer. Also, aluminum-lithium parts can be fabricated using current machinery, and new equipment and techniques that require huge investments are not necessary.

Aluminum-lithium alloys do have some disadvantages. For one thing, they are roughly two to three times more expensive than ordinary aluminum, due in part to the high cost of lithium. This means that lithium scrap must be salvaged, a labor-intensive process. Even so, aluminum-lithium components could cost considerably less than parts made of plastic composites, at least on a per-pound basis.

Other aluminum alloys are being developed to operate at much higher temperatures, even up to the temperature ranges that now require the use of titanium. Ordinary aluminum starts to lose its strength at 250–300 degrees Fahrenheit. Experts believe they can make aluminum-based materials that can function at temperatures of up to 650 degrees Fahrenheit. These aluminum alloys could be about 15-percent lighter and as much as 65-percent cheaper than equivalent titanium parts.

Superalloys

High-temperature applications are the focus of superalloy research. Many advanced superalloys gain their characteristics by combining basic materials like titanium, aluminum, beryllium, nickel, cobalt, molybdenum, and niobium. Titanium-aluminide, one such superalloy, will increase the current operating range of titanium from 950 degrees Fahrenheit to about 1500 degrees Fahrenheit, but without an increase in weight. Another metallic material gaining popularity is beryllium, one of the lightest structural materials known. Beryllium has excellent heat-dissipating properties, making it useful for such applications as braking systems on high-performance aircraft.

Great strides are being made in the way metallic materials are turned into useful components. Time-honored metalworking processes such as forging, casting, extruding, rolling, and machining are taking on new dimensions, and entirely new processes are being created because of the unique characteristics of advance metallic materials and superalloys.

One new metal processing technique is rapid solidification. Here, molten metals are cooled at tremendously rapid rates, as high as one-million-degrees per second. This results in very homogeneous materials because rather than large crystals being formed, the short cooling time leads to small, uniformly distributed crystals. This results in solidified alloys which are much stronger and have higher melting points (Fig. 6-2).

Because of the potential of rapidly solidified alloys, techniques are being developed and refined to produce the finely powdered materials that are the end product of the process. These powders are then usually pressed or extruded into ingots or billets, or even the final part. One of the simplest procedures, called splat cooling, involves spraying hot, molten droplets onto a cold surface. A variation on this idea is to spray the fine droplets into an extremely cold, inert-gas environment. Another idea is to pour the molten material onto a rapidly spinning wheel which, through centrifugal action, breaks up the material into fine droplets and throws them into a cold-gas atmosphere for rapid cooling.

The powders can be formed into useful shapes in several ways. Powders can be heated and then forced or extruded under high pressure through a small hole, much like toothpaste out of a tube, to form billets to be machined into the desired parts. In the dynamic-compaction technique, the powder is placed in a die shaped like the final component. The powdered material is compacted using a gun barrel and a projectile. The projectile is "fired" down the barrel by ultrahigh gas pressure, or even by explosives, and the gas in front of the projectile compacts the powder in the die. The final part produced needs no further machining.

Rapid solidification can also be used to put high-temperature and high-strength coatings on components such as parts of gas-turbine engines. One

Fig. 6-2. Here the fine powders of alloys are being formed by the rapid-solidification technique. After being sprayed out in molten form, the alloys are cooled very, very rapidly.

technique is to use a plasma flame to heat the surface of the part. The part is then rapidly cooled to retain the desired characteristics in the thin coating.

Thin foil of intermetallic compounds like titanium-alumide with thicknesses on the order of 0.003 inches are combined in layers with silicon carbide fibers to create an intermetallic-matrix composite that can be used in vehicle structures and skins, or for turbine engine components that can operate at high temperatures. These foils can be produced by a combination of surface grinding and chemical milling of ignots, but up to 95 percent of the material is wasted making the foil very expensive. Texas Instruments has developed a new technique to produce foil using cold rolling of plasma sprayed sheets rather than using cast ingots, over-coming problems with the material's negligible ductility at low temperatures that prevented cold rolling in the past. Waste is reduced to about 50 percent and costs drop by about a third. A further improvement uses powder plasma sprayed onto a stainless-steel drum, then run repeatedly through a special isobaric (constant pressure) cold-rolling mill. This process reduces waste to about 20 percent and reduces costs to a few hundred dollars per pound. This is a considerable improvement considering that to produce the same foil in the past using cold rolling of an ingot it cost several thousand dollars per pound.

Another process, called laser glazing, uses a high-powered laser to heat the part, forming a thin layer of molten material on the surface. The actual cooling is done by heat conduction at rates as high as 10 million degrees-per-second. Laser glazing has the advantage of only affecting the surface; the material below retains its original properties. Also, rather complex parts can be made by applying many layers.

One technique that has been around for a while, but now shows great promise for the future, is called superplastic forming. Here the metal, in putty-like form, is worked vigorously to obtain the optimum grain structure. The material is then formed into the desired shape before it is heat-treated, rapidly cooled, and aged (Fig. 6-3).

In another technique, called diffusion bonding, parts are joined together under very high pressures and temperatures. As the name implies, joining occurs because the atoms of the parts being mated actually flow, or diffuse, across the solid boundary. Unlike welding, the materials being joined do not melt.

Fig. 6-3. Titanium parts being made by superplastic forming. Insulated suits are needed by the handlers because temperatures are as high as 1700 degrees Fahrenheit.

Ceramics

A few years ago ceramic materials like silicon-carbide, silicon-nitride, titanium-oxide, and alumina-zirconia were touted as the materials that could spell the end of the Metal Age, at least in lightweight jet engines that could run hotter, faster, and for more hours. Unfortunately, ceramics have not been as successful as most other advanced materials. Indeed, aerospace applications were pretty much limited to the Space Shuttle's heat shields and to cutting tools for machining the superalloys. Ceramics are extremely brittle and completely lack ductility. Ceramic parts have to match perfectly or they crack, if the parts can be fabricated at all. Machining is time-consuming and their extreme hardness dulled cutting tools rapidly. Machining also introduced tiny cracks which eventually grew into flaws that cause the ceramic part to fail. Molding of ceramic parts requires high temperatures, high pressures, and much time.

However, there have been some breakthroughs that still make ceramics a material for the future. One of these is the development of ceramics with ultrafine grains. While ceramics (like metals) might look smooth, in reality they are composed of very tiny adjoining islands of matter, or grains. As the grain size shrinks from micron to nanometer size, important things happen to the properties of ceramic materials. For instance, they become more fracture resistant and more ductile. Scientists say that, like metallic materials, ceramics can exhibit superplasticity—that is they can be elongated to many times their original length. Superplastic ceramics can be stretched and stamped into a mold, or perhaps rolled into thin sheets. Besides being more ductile and maleable, nanophase ceramics should also be much tougher.

COMPOSITES

Like the transistor, microchip, and laser, composite materials are one of the great technological advances of the last half of the twentieth century, and they will play a major role in making future aircraft designs possible.

Materials that are combinations of two or more organic or inorganic components are given the generic name "composite." One material serves as a matrix, while the other serves as a reinforcement (in the form of continuous fibers dispersed in the matrix in appropriate patterns). The matrix material bonds the reinforcing fibers together as well as transferring loads between the fibers. The matrix can also provide the composite material with other desired properties, such as resistance to high temperature, radar absorption, or immunity to corrosion.

What makes composite materials so attractive is that the composite designer can blend the fibers and matrix into a new material with new and better properties than those of the individual constituent materials. For example, fragile glass, when spun into fine fiber, has six times the tensile strength of ordinary steel. But this extraordinary strength can only be taken advantage of if the glass fibers are embedded in a plastic-type material that is ductile. This allows the composite to be formed into a finished component that can be subjected to bending and twisting stresses. Often composites can be optimized for strength, stiffness or other mechanical properties merely by changing the orientation of the fibers. The use of composites essentially allows engineers to reverse the design process. In the past, they had to design around available materials. Now, they can specifically design the composite materials to meet the needs of the job. Finally, composite materials usually have very low density to go along with their high strength or other mechanical properties, making them ideal for aerospace applications where low weight is so important.

Organic composite structural laminates are made up of stacks of oriented thin lamina made of high-strength, high-modulus, low-density reinforcing fibers embedded in a resin matrix. The layers of lamina are permanently joined together under heat and pressure. The fiber part of the composite can be of such materials as fiberglass, boron, silicon carbide, or Kevlar and can range in size from small particles to long continuous threads. While particles and short fibers are randomly dispersed in the matrix material, long fibers can be woven, knitted, or braided into the final shape of the component. Then they can be embedded in the matrix material to obtain the desired characteristics. Indeed, composite material designers are now borrowing advanced techniques like fabric stitching, braiding and three-dimensional weaving from the textile industry.

The matrix can be either a thermosetting material, such as epoxy, bismaleimide or polyimide, or a thermoplastic material. Thermosetting plastics cannot be remelted after they have cooled in the forming process. The earliest, and most commonly used composite is fiberglass, made by imbedding glass in a thermoset plastic matrix. By contrast, thermoplastics can be reheated and reformed over and over again.

Fiberglass was first widely used in the 1950s for boats and automobiles (e.g., the Chevrolet Corvette). The Boeing 707, the first airliner to use composites, was only about two percent fiberglass. But the aircraft industry soon saw the advantages of these man-made materials. For example, composites account for about 10 percent of the Boeing 777's structural weight while the F-22 Advanced Tactical Fighter probably uses composites for over 30 percent of its structure. As a point of history, the British Aerospace-McDonnell Douglas AV-8B Harrier, which has been flying since 1981, was the first high-performance combat aircraft to make extensive use of composites, with over 25 percent of its structure made of composite materials (Fig. 6-4). Some experts predict that as much as 65 percent of twenty-first-century aircraft will be made of advanced composites, probably a mix of both thermoset and thermoplastic materials, providing they can be produced cost effectively.

Thermoplastics were introduced only a few years ago for aerospace applications, but probably show the greatest promise for the future. Thermoplastics offer increased toughness and durability over thermosets, and they are especially resistant to low-energy impact such as when a tool is dropped on a component. From a manufacturing viewpoint, thermoplastics can potentially reduce fabrication time. These materials only need to be heated enough to soften them so they can be formed and then cooled to a hardened state. Freezer storage is not needed.

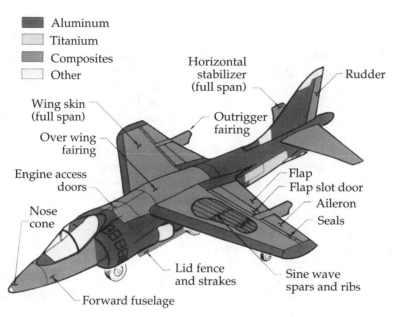

Legend:
- Aluminum
- Titanium
- Composites
- Other

Horizontal stabilizer (full span)

Rudder

Wing skin (full span)

Outrigger fairing

Over wing fairing

Engine access doors

Flap

Flap slot door

Aileron

Seals

Nose cone

Lid fence and strakes

Sine wave spars and ribs

Forward fuselage

Fig. 6-4. The AV-8 Harrier was the first high-performance military aircraft to use composite material extensively. McDonnell Douglas.

Thermoplastic materials are of great interest today because they are recyclable, and this interest will grow as resources become scarcer. New industries will spring up to recycle everything from thermoplastic beverage bottles and packing containers to automobile bodies and aircraft fuselages. Fortunately, progress is also being made in developing techniques to recycle thermoset composites as well.

One of the biggest drawbacks of composites is their higher cost compared to conventional metals, partly because they are more labor intensive to produce. Composite fabrication techniques like filament winding, press forming, pultrusion, and adhesive bonding are quite complex and require large initial investments (Fig. 6-5). Fortunately, much is being done to reduce costs, including using more automation, minimizing part counts, and introducing more commonality in designing parts.

Although composites are usually made from petroleum-based chemicals, their petroleum usage is very small compared to total petroleum usage. They also represent a very efficient use of petroleum and result in high-priced items, meaning large payoffs for the resources invested. The composite industry would probably be the last to be drastically affected by future petroleum shortages.

Although matrix material is most usually thought of as some type of plastic, there are also metal-matrix composites (MMC) that combine the advantages of both metallic and non-metallic technology. Here, fibers made of such materials as silicon carbide, boron carbide, alumina, and graphite are combined with a metal matrix made of aluminum or magnesium. Again, the fibers can range anywhere in

Fig. 6-5. Giant autoclaves are used to cure the advanced thermoplastic-resin panels used on the F-22.

size from microscopic particles to continuous strands running the length of the finished structure. The resulting MMC is stronger, stiffer, and lighter than pure aluminum or magnesium, and its other properties, such as thermal conductivity, are enhanced. For example, when ceramic particles such as silicon carbide or alumina are added to aluminum for a total content of 10 to 20 percent, stiffness is boosted by 30 to 50 percent and strength is increased by 15 to 40 percent over the same basic metal at the same density. Structural benefits include about a 25-percent lighter weight and a material that keeps its strength up to around 500 degrees Fahrenheit.

Much more materials research is underway in university, government, and industrial laboratories today that will make the aircraft of the twenty-first century possible. But it takes a long time, sometimes as much as 15 or 20 years, from the time a new material or process is "invented" until it is actually used on a production aircraft. During this interim time, extensive testing is required to fully understand its properties, and fabrication techniques have to be perfected to allow economical manufacture.

FACTORIES OF THE FUTURE

A revolution is already underway in the way aircraft are designed and built, with the main objective being to reduce the spiraling costs of new aircraft. Technologies like the high-speed computer, robotics, artificial intelligence, and machine vision are the keystones of this revolution.

In the aerospace industry, most labor costs result not from the people who do the hands-on work in building aircraft, but from the people who charge their time to overhead. These include the tiers of managers, supervisors, schedulers, quality control experts, and a myriad of other people currently needed to keep the aircraft production line going. Because of the complexity of new aerospace systems, the demands of the customer (often the government), and low production rates for most systems, overhead dominates the total cost of building a new airplane. Depending on the aircraft, overhead costs can account for up to 70 percent of the total price of an airplane.

Fortunately, the computer does its best work on exactly these overhead tasks. Computers are taking over many of the tasks done previously by roomfuls of people, including filing and retrieving data, scheduling material and machine use, making design changes, and monitoring the quality of the product. One example of this emerging capability is the 2D/3D electronic product definition computer system used by the USAF, Northrop, and its subcontractors to keep track of the myriad of details involved in the development and manufacture of the B-2 (see Chapter 7).

Today, there is great interest in computer-integrated manufacturing (CIM), which integrates the now developed concepts of CAD, CAM, and CAE (computer-aided engineering).

CAD replaces the traditional drafting board for the design engineer (Fig. 6-6). The designing of an aircraft part, or even an entire airplane, is done on a computer screen. The Boeing 777 jetliner is the first airplane to be 100 percent digitally designed. CAD has progressed to the point where the designer can see the work in multicolors and rotate it to any position. The designer can "preassemble" all the parts on the computer screen to see how parts will fit together and see any potential misalignments, reducing the need to build expensive mockups. Changes to the design can be made via keyboard or light pen (Fig. 6-7).

When the designer is satisfied with his work, he or she gives the system a command to print the drawings so they can be sent to the factory. Better yet, they can be converted to computer instructions that are read directly by the machines of a CAM system, like robotic machines that might cut metal or paint parts (Fig. 6-8). In the CIM triad, computers and software do the multitude of engineering calculations to ensure that a part or an entire airplane will do its job over its intended life. Included in this CAE concept are engineering analyses that range from aerodynamic and thermodynamic calculations to stress and fatigue computations.

The three component systems of CIM have to talk to one another using common computer languages and common nomenclature. Artificial intelligence will undoubtedly play an even more important role in the CIM system; expert and neural AI techniques will weigh alternatives and automatically make logical choices.

CIM offers many advantages, including ease of redesign. This is very important in the aerospace industry where parts and aircraft are made in relatively small quantities. While it might take a machinist hours to set up his milling machine or lathe to make just one new part, a single CAM machine can be reprogrammed very rapidly to make a multitude of different parts.

Fig. 6-6. Aircraft parts are designed using computer-aided design programs. The final design can then be converted to models made of photosensitive liquid plastic hardened by laser light, using a technique called stereolithography. The models allow designers to see their designs in three-dimensions and is quicker and much less expensive than building handcrafted models.

CIM cannot only instruct the CAM machines to make the part, but also instruct material handling equipment to get the raw material from the storage area and route it through the entire manufacturing process. The material handling aspect is a very important part of the increasingly favored "just-in-time" inventory philosophy, in which only enough inventory is maintained to keep production going and materials are delivered to work stations only as needed. While the concept reduces inventory costs and facilities tremendously, it does require very sophisticated computer control.

Quality control is another area that will see tremendous automation. In the future, parts will not have to be inspected by humans. Sensors located within machines, in furnaces, and on assembly lines will immediately sense when things are not perfect. Computers, using artificial-intelligence techniques, will not only stop production, but will analyze the situation and make the necessary corrections. Inspection techniques will rely heavily on machine-vision technology to "look at"

Fig. 6-7. The entire Boeing 777 is completely "pre-assembled" on the computer screen before the first prototype aircraft is built. Here an engineer is designing a fuselage section that includes the routing of all the required systems. He is able to see if pieces fit together and if there are problems before the final design is manufactured.

objects and digitize what is "seen" into the language a computer can understand. Other fault-sensing techniques will use ultrasonics, infrared sensors, and lasers.

Naturally, robots will populate the automated aircraft factories of the future. Because they are not subject to the human frailties of boredom, fatigue, or an "off day," robots can be programmed to unerringly produce parts to exact specifications. They also work well in toxic or hazardous environments.

Although automated aircraft manufacturing will require fewer people, the human element will still be the most important ingredient. People will, however, be removed from the tedious jobs. Instead, they will be required to write the software, design the programs and processes, monitor the overall system, be the "experts" for the artificial-intelligence systems, and finally, maintain and repair all the sophisticated equipment.

Automated equipment will not only be found in factories producing new aircraft, but also in facilities that are "recycling" old aircraft to update them with the latest technology. Development is already underway on automated equipment that can strip paint, remove rivets, and duplicate new parts from old parts (when

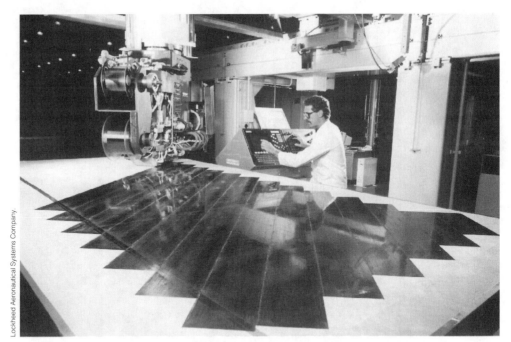

Fig. 6-8. A computer-aided machine cutting thermoplastic resin in strips.

the original engineering drawings are no longer available). As stated earlier, there will be intense interest in modernizing the aircraft in use today so they can perform first-line service well into the twenty-first century.

7

Military aircraft

THE TWENTY-FIRST CENTURY will see a few completely new aircraft designs among the many familiar airplanes that are flying with the military today. Only when current aircraft, or modified and updated versions of current aircraft, can no longer do the job will scarce national resources be committed to the development of entirely new systems.

Because of the long lead time between design conception and actual fielding of an operational unit (today this can take as long as a decade or more), the designs now in the planning and development stages will be the mainline systems of the next century. That is, of course, if they can pass the myriad of approvals necessary and do not get cancelled along the way.

Every major new military system represents a significant portion of the national defense budget, so there will be absolutely no room for "white elephants" in future military inventories. Unfortunately, this environment could produce conservative designs; the military and the manufacturers may shy away from promising technologies that have even the slightest bit of risk. Technologies will have to be well proven before they are incorporated in a new design.

F-117A STEALTH FIGHTER

The story of stealth technology is probably best told by looking at the Lockheed F-117A fighter. This sharply-angled, black fighter made its highly successful debut during the Persian Gulf War with 56 aircraft performing 1270 missions virtually undetected by the enemy (Fig. 7-1). Like many other successful-advanced technology, top-secret projects, the F-117A was developed at the famed Lockheed Advanced Development Company, better known as the "Skunk Works," under the leadership of the legendary aircraft designer and developer, Kelly Johnson.

The development of the world's first stealth aircraft started in the 1970s to counter a rapidly improving Soviet air defense threat that included early warning capability backed up with manned interceptors and surface-to-air missiles. The USSR's Mainstay airborne warning and control system used an early warning radar mounted on top of the IL-76 transport plane. MiG-29 Fulcrum and Su-27 Flanker fighters, with their radar and missiles, were quite capable of detecting and

Fig. 7-1. The F-117A is on the left, the YF-22 is on the right. Note some features, like the tail design, have been carried over to the Air Force's Advanced Tactical Fighter. The tail's oblique orientation on the F-117A not only reflects radar energy, but also shields the F-117A's hot exhaust from detection by "look down, shoot down" capability.

destroying low-flying aircraft, even when masked by radar returns from ground clutter. SA-10 ground-to-air missiles could engage several low-flying aircraft and cruise missiles at the same time, and Soviet phased-array radar could detect and track multiple air targets. These weapons posed a threat over the skies of the Soviet Union and worldwide for years, even decades to come, as much of this advanced military technology was exported to other nations around the world.

The stealth technology base for the F-117A can be traced back to the late 1950s when initial low-observable ideas were first investigated. These ideas were needed to improve the Lockheed U-2's ability to evade detection during its reconnaissance flights over the Soviet Union. Incidently, the A-12 and SR-71 incorporated initial low-observable technology. However, the F-117A would be the first aircraft that would be designed foremost to be stealthy, and to put much of the stealth technology that had only been tried on computers and in laboratories into practice.

The first F-117 prototype flew in 1981, the first production aircraft was delivered to the USAF in 1982, and the F-117A became operational in 1983. In all, 59 F-117As were built at a rate of less than eight per year. The 59 aircraft each cost almost $43 million, or about $112 million if all the associated research, development, and testing costs for the entire program are factored in.

In order to make the F-117A all but impossible to detect under all conditions in which it would have to operate, the engineers had to be concerned with no less than seven different types of observables: radar, infrared, visible, contrails, smoke, acoustic, and electromagnetic. An important part of the stealth equation is time. Flying at a top speed of just under 650 MPH, the F-117A depends on being "invisible" to enemy detectors until it is too late for the enemy to take action. Thus, while the F-117A is not completely invisible, with all the different types of low-observable technology working together they significantly reduce the time for detection, tracking, and finally targeting.

There are basically two ways to avoid radar detection. Either reflect the radar return so it does not return back to the receiver, or use radar absorbing materials

(RAM) to absorb the radar energy so there is no return. The F-117A uses both techniques.

For starters, the F-117A contains the greatest collection of straight lines and flat surfaces seen in the skies since the biplane fighters of World War I. The aircraft has many edges, surface intersections, and sawtooth panels that run in parallel directions to change the direction of the reflected radar beam. You will not find sharp corners at near 90-degree angles where the surfaces of the plane meet. Sharp corners are great radar reflectors and are often placed purposely on small objects to increase their radar reflections. For example, the "V" tail surfaces of the F-117A are mounted at oblique angles with respect to the fuselage. There are also no gaps, cracks, or openings on the fuselage; these are great radar cross-section (RCS) enhancers. While not a supersonic aircraft which requires swept wings, the F-117A's leading edges are swept at up to 67 degrees for another reason, to make sure that radar coming from the dead-ahead direction, the direction where radar defenses would mostly likely be located, is reflected well off to the sides of the aircraft. Incidently, supersonic capability would have defeated many of the F-117A's stealth characteristics, or at least made them very difficult to achieve. For example, sonic booms would have been a dead giveaway, and an afterburner would have lead to a much larger, heavier, and thirstier aircraft. Aerodynamic heating, a by-product of supersonic speeds, would have brought severe infrared and radar detection problems.

Radar energy that is not reflected is absorbed by the RAM that is incorporated at strategic locations on the aircraft. Some of the applications used in the F-117A include a ferrite-type RAM. This material absorbs radar energy by converting it to heat. When the radar beam strikes the ferrite, the energy makes the molecules oscillate creating thermal energy. Attention to detail is a key element of a successful stealth design. For example, much effort was devoted to reduce the radar cross-section (RCS) of the four pitot tubes. When a RAM did not work, the engineers took three years to find another solution, probably an electrically conducting plastic.

The black paint scheme was used to limit visibility at night. For daytime flight, it could be painted virtually any color. The main structure of the aircraft is aluminum, not composites which were very expensive when the F-117A was designed. However, composite components can be retrofitted as is the case with the composite rudders that have already been installed.

Engines give stealth designers some real headaches. Up front inlets, great reflectors for radar, are needed to supply intake air to the engines. Because of the F-117A's relatively small size, the twin General Electric F404 non-afterburning turbojet engines could not be buried in the wings. Instead, the engines lie behind the rectangular intakes masked by RAM-lined square tubes located in the inlet that, when viewed head on, look like a grid. Any radar energy travelling through the "tunnels", that are several inches long, is not reflected enough to be useful in detection. A mechanical wiper can be extended from below the grid to clear away any ice that might hinder air intake. At the other end, the jet engine produces a red hot cylindrical exhaust plume that can be detected by infrared (IR) sensors or used for "homing" by heat seeking missiles. The F-117A incorporates nickel-alloy honeycomb tailpipes that flatten the exhaust, drastically reducing the IR signature as well as cooling the exhaust. The wide nozzles mix cooling air from the engine fan with the hot exhaust from the core of the engine. Vertical vanes in the exhaust nozzles and an extended upturned lower lip of the tail hide the hot engine parts and the spinning turbine blades. Non-afterburning engines provide a relatively low-noise signature. The F-117A's shape, engine, inlet, and nozzles were selected to make it as quiet as possible. The General Electric F404 engines were built to be smokeless, and fuel additives are used to reduce contrails, or vapor trails.

Considering that this is a single-seat aircraft, it is rather large, actually bigger than the F-16. All its fuel and weapons have to be carried internally to keep it stealthy. Hanging fuel tanks and ordnance on the wings and fuselage usually result in a very "dirty" radar signature. And compared to most sleek, modern fighters, the F-117A looks a bit ungainly, even unstable. However, modern fly-by-wire control systems and high-speed computers that make dozens of corrections every second for instabilities, could allow a barn door, let alone the F-117A, to remain stable, provided enough engine thrust was available.

Since on-board radar can be detected by the enemy, the F-117A uses only a small, downward-pointing radar altimeter that is used for low-altitude terrain following. The pilot navigates using an inertial navigation system assisted by an infrared imager that provides an image of the scene ahead on a cathode ray tube (CRT) and pilot head-up display (HUD). There is also a laser with the IR system that is used to designate targets for the two laser-guided 2000 pound "smart bombs."

B-2 STEALTH BOMBER

While equally stealthy, the B-2 bomber, because of its size, intended missions, and more recent design, incorporates some different techniques to achieve its low observable characteristics. For example, the B-2's continuously curved surfaces on the basic flying wing planform represents the latest thought in stealth technology for deflecting incident radar beams. By comparison, at the time the F-117A was designed, stealth technology and manufacturing capability dictated that flat, two dimensional shapes be used to achieve low radar visibility. Even so, the B-2 still uses parallel edges in many locations, such as on engine inlets and wing trailing edges (Fig. 7-2).

The F-117A is designed for shorter range attack missions; the B-2 is a long-range penetrator with up to 6000 nautical miles possible unrefueled and an additional 4000 nautical miles with a midair fillup. Compared to F-117A's somewhat modest payload capability, the 376,000 pound B-2 can carry more than 20 tons of nuclear or non-nuclear weapons. Like the F-117A, everything is carried internally. The B-2 normally flies with a two-person crew, but there is room for a third-crewmember if future missions require him. Long-range flight mandates that more attention be paid to drag reduction to prevent unacceptable fuel consumption. This is the reason the B-2 does not have a drag increasing vertical tail. A sophisticated computerized fly-by-wire electronic flight control system is the reason the B-2 flies, and flies well.

Fig. 7-2. Viewed from the top, you can see how the hot exhaust from the B-2's twin engines is masked by the fuselage.

Because of its larger size, the B-2 bomber's 19,000 pound thrust-class engines can be buried in a deep wing. Also the radar-reflective engine faces can be somewhat hidden by a curved engine duct. Likewise, the hottest portion of the exhaust plume is masked by the fuselage. Incidently, the B-2's General Electric F118-GE-100 engines were derived from engines used in the B-1B, F-14 and F-16.

The B-2 earns the title Advanced Technology Bomber (ATB) because of its many innovations, including the extensive use of advanced composite materials that combine high strength and stiffness with low radar observability. These materials which often bring manufacturing challenges, plus the need for the extremely tight tolerances demanded by stealth requirements, led to many breakthroughs in manufacturing technology. For example, the accuracy of the 172 foot wingspan is held to within 0.25 inches (Fig. 7-3). In fact, the B-2 industrial team that included more than 4000 companies large and small across the nation developed nearly 900 new materials and process techniques. Many of these developments and B-2 processing equipment are now being transferred to U.S. industry for both military and commercial applications (Fig. 7-4).

For example, B-2 manufacturing requirements led to the development of a high-speed milling machine that operates at a spindle speed of 100,000 RPM,

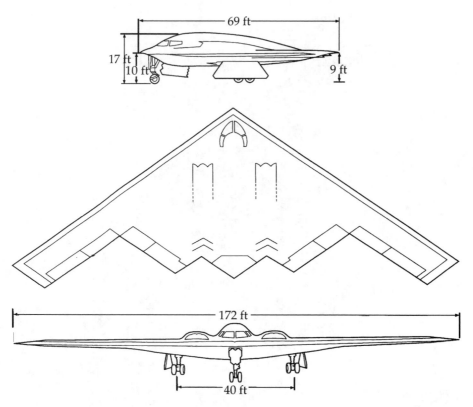

Fig. 7-3. The tailless B-2 bomber has a wingspan that is much greater than its overall length. Without its advanced fly-by-wire control system the B-2 would be difficult if not impossible to fly. Northrop Corporation.

about three to five times the speed of standard high-speed milling and routing machines. This results in reduced processing time while increasing quality. The Adaptive Control Drilling System is a computer-controlled drill that senses the hardness of different materials such as aluminum, titanium, graphite composite, etc., automatically adjusting the bit speed for the particular material. It also automatically retracts the bit periodically to allow cuttings to be removed. Besides reducing average cutting time by one-half, it also extends the amount of time between having to resharpen drills by four. A composite materials cutter using ultrasonic energy cuts materials to the required shape three times faster with a threefold increase in precision. An automated tape lamination process is up to 60 times faster than conventional methods. These automated machines allow fabrication of composite components into some of the largest primary structures ever made with the tolerances needed to be stealthy.

Computer technology played a major role in making the B-2 happen. No prototype was ever built. The first B-2 was the first all-new aircraft to be built on production, not prototype, tooling. The plane first shown to the public was the first one ever built. It flew for the first time on July 17, 1989. The plane was manufac-

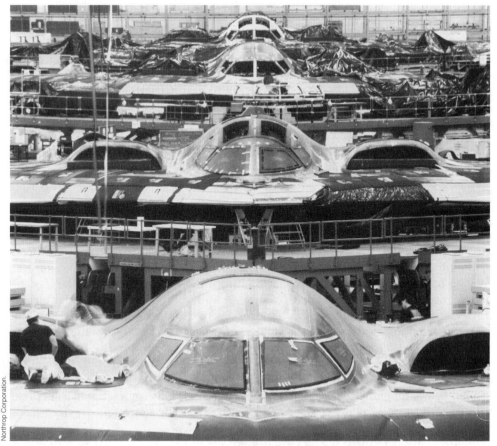

Fig. 7-4. Building the B-2 has brought advances in manufacturing technologies that rivals the advances made in low observables.

Northrop Corporation.

tured, tested, and serviced on a computer screen before it was built. This is not to say that the B-2's were not tested. Indeed, the B-2's underwent an unparalleled 800,000 hours of testing, including 24,000 hours of wind tunnel time, more than any other subsonic aircraft. Northrop and USAF pilots accumulated some 6000 hours in B-2's advanced flight and mission simulators.

The complete B-2 design is kept up-to-date in the Northrop 2D/3D (2-dimensional/3-dimensional) electronic product definition computer system, a form of an advanced computer-aided design and manufacturing (CAD/CAM) system. The system maintains all the details in easily understood 3D graphics and with accuracies measured in billionths of an inch. The system defines each one of the B-2's parts in three dimensions and stores and communicates the information in data form. Major subcontractors and the Air Force are tied into the system to make sure everyone is working from the same up-to-date information. Logistics, maintenance, and support organizations also use the system and Northrop is licensing it for commercial use by other companies.

The B-2 is also kept flying using computer technology. The On-Board Test System (OBTS) automatically tests, reports, and records data from thousands of B-2 components and systems before, during, and after each flight. After each flight the OBTS printer located in the cockpit prints out a listing of faulty line replaceable units (LRU) or other items that need servicing. It also gives status of consummables like hydraulic fluid and engine oil. Using the OBTS printout, the crew chief can schedule immediate servicing and repairs so that the B-2 can be ready for the next flight. Instead of referring to bookshelves filled with technical orders and instructions, the technicians go out to the airplane with an Improved Technical Data System (ITDS), which looks like a laptop computer, that contains all the information needed to perform the job.

HYPERSONIC RECONNAISSANCE AIRCRAFT

In 1990, the USAF retired the SR-71 Blackbird high-speed reconnaissance aircraft after more than twenty years of service, but with no obvious replacement for it. While much high quality reconnaissance is accomplished by our "spy" satellites, they do have some limitations because their locations are governed by the laws of orbital mechanics. While satellites can fly over and photograph a particular target, it may take 24 hours before it is in the proper orbit to do so. By contrast, spy aircraft can provide "on-demand" reconnaissance. This not only allows the most current intelligence data, but also fly overs can be made under the best lighting conditions. Unlike a satellite whose position can easily be predicted, the enemy cannot predict when aircraft will fly over and thus they are not sure when to hide what they do not want to be seen. For these reasons, it was somewhat of a surprise that there was no obvious replacement for the SR-71 to complement reconnaissance satellites.

Now it appears there was a replacement for the SR-71, but it was kept super secret, just as it should be for reconnaissance missions. At this writing, a hypersonic reconnaissance aircraft called the Aurora, is postulated by several expert "aviation watchers" to be the replacement for the SR-71. While details are still rather sketchy, the Aurora probably has a cruising speed as high as Mach 8, or over 5000 MPH, and can fly at altitudes of 100,000 to 130,000 feet. This speed means the aircraft could reach any point in the world in under three hours. The Aurora is thought to be propelled by a combined-cycle engine that uses liquid methane. The fuel, stored at cryogenic temperature, could be used as a heat sink and circulated for cooling the hottest parts of the structure, such as the nose, leading edges of the planform, and inlet, before being delivered to the engines. The lifting body is around 90-feet long with a sweep back of 75 degrees to give a span of just under 50 feet for its triangular planform.

ADVANCED TACTICAL FIGHTER (ATF)

The Advanced Tactical Fighter was developed as the eventual replacement for the McDonnell-Douglas F-15 to become the U.S. Air Force's mainline air-superiority fighter for well into the twenty-first century. When initiated in the mid-1980s, the intended threat included two first-class Soviet fighters, the SU-27 Flanker and MiG-29 Fulcrum, plus much more capable Soviet airborne and ground-based radar systems and missiles. The ATF will not be in operation until at least the year 2000, when the F-15 will be about three-decades old. Since the task was too costly for any single company to undertake, two teams of aerospace contractors developed prototypes. Lockheed, Boeing, and General Dynamics produced the YF-22, while the competing YF-23 model was developed by the Northrop and McDonnell-Douglas team. After extensive testing, the Air Force chose the YF-22 over the YF-23. The YF-22 made its first flight on September 29, 1990 (Fig. 7-5).

Fig. 7-5. One of the two YF-22 Advanced Tactical Fighter prototypes shown in flight over the Mojave Desert during flight tests at Edwards AFB, California.

Three key objectives that had to be balanced throughout the F-22's design: the ATF had to be stealthy; it had to provide "supercruise," that is fly supersonically without the use of an afterburner; and it had to provide a "first look, first kill" capability. Low observables dictated that stores and fuel be carried internally. By not using an afterburner, the aircraft's infrared signature is much smaller while using much less fuel. Also high on the priority list was the requirement to be highly maneuverable and capable of penetrating and fighting in high-threat environments worldwide. The F-22 is designed to gain and maintain air superiority over the battlefields of the foreseeable future.

The F-22 has a wingspan of 43 feet and is 64.2-feet long and, thus, is only slightly larger than the F-15. Unlike the F-15 which flies with ordnance and fuel tanks hung onto its frame, the F-22 is designed to carry its weapons and fuel entirely inside to reduce aerodynamic drag and radar reflections. A typical weapons load might include four AIM-120 Advanced Medium-Range Air-to-Air Missiles (AMRAAM) and a couple of AIM-9 Sidewinder heat seeking missiles. Although it carries fewer weapons than the F-15, the F-22's lesser capacity is more than compensated for by its advanced, integrated avionics which are designed to deploy the ordnance more effectively. Ordnance could be hung under the F-22's wings, but this increases aerodynamic drag and stealthiness is greatly reduced. There is an internally mounted 20 millimeter Gatling gun, modified from the gun used on the F-15 and F-16, which is capable of firing at 4000 or 6000 rounds-per-minute.

The F-22 includes high angle-of-attack flight and maneuver capability. Fighter pilots can use the F-22's superior agility to control the setup for the fight, maintain the offensive, and accomplish a quick kill. The F-22's maneuverability comes from an advanced flight control system that is fully integrated with propulsion controls including nozzle-vectoring of engine thrust. Thrust vectoring provides the control power for maneuvering at high angles of attack, while minimizing the weight of the airplane.

Lockheed, with its vast stealth technology experience from low-observable aircraft like the SR-71 and F-117A, incorporated much of the latest techniques to mask all telltale sources of electromagnetic or infrared radiation on the F-22. Like the F-117A, the F-22 has somewhat angular lines to reflect radar waves and to keep them from returning to the enemy's radar receivers. The outward-canted vertical tail both directs radar returns away from the radar source and provides additional directional stability at high angles of attack. Like the B-2, appropriately curved surfaces are also used for radar deflection.

Engine inlets, ducts, and exhaust systems were designed for low observability with minimum compromise of the propulsion system efficiency and the ability to provide supercruise and thrust-vectoring capabilities. For instance, the inlet/duct design provides low observables without any special engine face designs that can severely decrease the engine's ability to produce thrust. At the other end, trailing edge orientation and surface discontinuity techniques reduce the observability of the exhaust plume. Great attention to detail was given to make sure apertures for the fire control radar and other avionics plus crew station windows were stealthy without compromising their intended functions (Fig. 7-6).

Stealthiness is just one part of the "first look, first kill capability," equation that results in the ability to see an adversary and fire weapons before the opponent sees you. Advanced avionics and sensors are another important part of the equation. The integrated avionics are ready to handle twenty-first century threats while also reducing pilot stress and workload. Workload reduction is enhanced by providing the pilot with only the level of information needed for decision making and task management without burdening him with details about system operation.

The F-22's electronics include a fully-integrated computing core architecture and Very High-Speed Integrated Circuits (VHSIC) technology. Easily replaced modules are used instead of the traditional black boxes. Data transfer is via fiber-optics. Software is programmed in the military-standard Ada language.

The F-22 incorporates modular avionics, meaning the avionics are partitioned into highly-integrated, common, and modular building blocks. A common-integrated processor (CIP), designed by Hughes Electronics, represents an innovative application of modular avionics. The CIP does all the signal processing, data processing, digital input/output, and data storage functions using a single integrated hardware and software design. The system's total signal and data processing capability is massive—over 350 million-instructions-per-second for general purpose processing and 9 billion-operations-per-second of parallel programmable signal processing. With a digital flight-control system (DFCS), the F-22 pilot essentially

Fig. 7-6. The F-22 pilot has an outstanding view from the cockpit. The one piece canopy is another example of advances in materials.

needs only to command the aircraft to perform maneuvers. The DFCS does the fine-tuning and performs the maneuvers by electronically commanding actuators which move control surfaces such as the horizontal tail, rudder, flaperons, ailerons, leading edge flaps, and speed brakes. Fly-by-wire controls allow maximum flexibility for tailoring flying qualities. For example, leading edge and trailing edge flaps can be programmed for near-optimum wing camber for good takeoff and landing characteristics, or for high-speed maneuvering. Fail-operative redundancy contributes to a high probability of mission completion, not to mention a high degree of safety. This "smart" DFCS computer allows automatic structural load limiting, ensuring maneuvers do not exceed any of the aircraft's structural limits throughout the flight envelope. Software changes in the flight control computer can be made without removing the computer from the aircraft.

The F-22's "fighter pilot's cockpit" is designed to carry out the "first-look, first-kill" philosophy. Proven technology was used wherever possible to reduce risk while retaining the ability to adapt the cockpit for changing requirements and insertion of new technologies, such as helmet-mounted displays, three-dimensional displays, and voice control. Routine cockpit functions are automated to reduce pilot workload while still providing the pilot with full insight into system operation and functions.

The pilot sits behind six flat-panel, liquid crystal displays (LCD). Compared to CRTs, LCDs require less space, give better performance, are lighter, and use less electric power. An 8 × 8-inch primary multifunction display shows tactical information, while two 6 × 6-inch secondary multifunction displays show attack and defensive information. All three are full-color liquid crystal displays. The pilot can select the level of detail to be displayed and symbol shapes convey information on threats and targets, system status, and friendly, enemy, and unknown forces, as well as indicating the relative importance of the information. Two more displays are for aircraft subsystem, stores management, and checklist presentation. Finally, an up-front control/display is used for communication, navigation, and identification data as well as caution, warning, and advisory messages. Information displays eliminate unwanted and unneeded warning lights. The pilot is not only informed of what is wrong, but of what are the required corrective actions. There is a holographic head-up display (HUD) that presents primary flight reference, weapon-aiming and release information to the pilot as he looks out the window, rather than having the pilot divert his/her attention inside the cockpit for this data (Fig. 7-7).

Lockheed Aeronautical Systems Co.

Fig. 7-7. Compared to the clutter found in many current sophisticated fighters, the F-22 displays and controls are easy to use. Note the sidestick control at the pilot's right side.

The hands-on throttle and stick scheme (HOTAS) allows pilots to carry out an attack from beyond-visual-range to a one-on-one dogfight without having to remove their hands from the stick or throttles to manage offensive and defensive sensors and weapons. The throttle control provides pilots with a single grip for controlling both engines.

While the prototype YF-22 used advanced composites for 23 percent of the aircraft by weight, production F-22s will use at least 35 percent, including both thermoset and thermoplastic materials. Applications include center fuselage ducts, frames and bulkheads, as well as wing and control surface edges fabricated of carbon/bisalemide materials. Thermoplastic unidirectional and fabric materials with embedded carbon fibers are used for fuselage and wing skins.

Several fighter development programs are underway in other parts of the world. Saab is working on its multirole, JAS 39 Gripen that will meet Sweden's, and perhaps other countries', needs for an interceptor, reconnaissance, and air-attack fighter well into the twenty-first century (Fig. 7-8). In France, Avions Marcel Dassault-Breguet Aviation is developing an advanced-technology, Mach 2+ fighter called the Rafale. The British, Spanish, Italians, and until recently, the Germans have been involved in the European Fighter Aircraft (EFA) program and Japan is developing its FS-X close support fighter.

Saab-Scania.

Fig. 7-8. The Saab JAS 39 Gripen is a multirole fighter that easily converts from interceptor to reconnaissance to air-attack fighter.

What about future fighters? They will incorporate those technologies that promise performance and operational improvements, but were not quite ready for aircraft like the ATF. The cockpits of twenty-first century fighters will use the results of expanding computer and electronics technology. If pilots think the HUD and HMD are great, wait until they see the three-dimensional spherical display that could replace them. This wraparound display is so advanced that engineers are not sure how they would make it work, but they want it because of the flexibility it offers, as well as the vast amount of information that can be presented in easy-to-understand form.

Other features of the future fighter will include active flight controls that will allow the fighter to fly through severe weather and violent maneuvering with minimum damage to the airframe. Self-repairing flight controls will be able to reconfigure an aircraft to continue to fly and fight, even if it loses control surfaces, such as its elevons or rudder.

The flight-control system could even sense when the pilot is unconscious because of combat injuries or a G-induced blackout. One technique for determining consciousness is to monitor the pilot's blinking. (When a person blacks out, his eyes do not automatically blink, and just before unconsciousness his gaze is fixed.) This might be combined with other monitoring techniques like watching for the drooping of the pilot's head, absence of a firm grip on the controls, or the loss of blood pressure pulse in the brain. Much research is currently underway to reduce the chances of the pilot blacking out in the super-agile fighters of the future. Solutions could include more sophisticated G-suits and even special drugs.

For propulsion, these advanced fighters could have engines with 1000 to 2000 parts, compared with the 15,000 to 20,000 parts used today. This drastic reduction could be achieved with advanced materials and manufacturing techniques that allow complex parts to be produced as a single unit with a minimum of machining and finishing. Weights will also be reduced so that engines will routinely achieve 20-to-1 thrust-to-weight ratios.

HELICOPTERS

Helicopters, the U.S. Army's most important aircraft, also use advanced technology to perform missions ranging from scouting and reconnaissance to evacuating casualties from the battlefield. The Army's latest helicopter, the RAH-66, is a good example of the advanced technology used in a rotary-winged aircraft.

The RAH-66, an armed reconnaissance/attack helicopter, was designed to meet the Army's new post-Cold War strategy requiring the ability to react to regional conflicts using fewer people. The strategy also requires self-deployable aircraft based in the continental United States. The letters RAH stand for reconnaissance/attack helicopter. Besides behind the-lines reconnaissance, the RAH-66 will be used in armed scout missions to seek and destroy enemy forces, and for designating targets with a laser for the artillery. Previously, teams of gunships and light-armed scouts were often deployed to perform these tasks, but the RAH-66 can perform them by itself. About 1300 RAH-66 Comanche helicopters could replace nearly 3000 AH-1 Cobra, OH-6 Cayuse, and OH-58 Kiowa helicopters that are tactically and technologically obsolete, as well as increasingly more expensive to maintain. The RAH-66 is the first Western helicopter designed with the capability to defend itself against other helicopters. Downed and injured crewmembers from other aircraft can be rescued and carried to safety in the open weapons bay. Like almost every Army chopper, the RAH-66 also carries the name of an Indian tribe, Comanche in this case (Fig. 7-9).

Fig. 7-9. The RAH-66 Comanche is the first helicopter designed from the start to be stealthy and is the Army's first all-composite helicopter.

The Comanche, designed by a Boeing-Sikorsky team, is the first helicopter designed to emphasize low observables. This plus speed and agility enables the Comanche to avoid detection. Like the F-22, the Comanche has a "first look, first shoot" capability. Much of the RAH-66's stealthiness comes from its ability to fly very low so its presence is masked by trees and the terrain. Once near the target, the Comanche pops up for a few seconds to perform its job, such as viewing a target, firing weapons, or designating the target with a laser. By flying low among the ground clutter, radar cross-section signature reduction is not quite as critical as with a fixed-wing aircraft. To reduce the exhaust's infrared signature, the hot exhaust first passes through the tail boom where it is mixed with cooler outside air before being dumped downward through slots along the bottom of the tail boom. Noise, a real problem with helicopters, has also been reduced. The RAH-66's rotor spins at a lower speed leading to lower tip speeds, and thus is quieter than other helicopters, even those with a lesser number of blades. The shrouded Fantail antitorque system is quieter than conventional tail rotors.

The RAH-66 is 47-feet long, including the 39-foot diameter, five-bladed main rotor. The fiberglass bearingless main rotor has 40 percent fewer parts than the articulated rotor systems used in most helicopters. With bearingless rotors, the blades are twisted rather than pivoted with hinges, as in articulated systems. The RAH-66 is powered by twin T801-HLT-800 LHTEC turboshaft engines built by General Motor's Allison Gas Turbine Division and Allied Signal's Garrett Engine Division. Rated at 925-shaft horsepower, the T-800's weigh just over 300 pounds each and are 30–50 percent more fuel-efficient than current turboshaft engines. The bottom line is outstanding maneuverability that includes a top dash speed of just over 200 MPH. It also includes the ability to make a snap turn in three seconds to dodge an enemy bullet and without having to perform a bank-to-turn maneuver. The landing gear is fully retractable to reduce drag.

The two man crew sits in tandem with the pilot up front. The low-workload cockpits are redundant including the versatile flat-screen displays so both crew-members can fly the aircraft. Since the RAH-66 is designed to operate in nuclear, biological, and chemical environments, the cockpit is sealed and pressurized. Navigation is via the Global Positioning System aided by a laser gyro and a Doppler radar system. A "God's eye view" 3-dimensional digital moving display based on the system used in the F-117A keeps the crew appraised of their overall situation. The crew uses their Standard Field-of-View Helmet-Mounted Displays to operate the two nose turrets that point sensors and weapons as well as to monitor system status without having to look down into the cockpit. Visors in the helmet-mounted displays protect eyes from laser damage. Sensors include the latest generation Martin Marietta focal-plane-array, forward-looking infrared (FLIR) night vision, and a target acquisition system with enhanced resolution and extended range compared to earlier systems. The Westinghouse Aided Target Detection and Classification System includes a data library to aid target recognition via algorithms that compare the FLIR image stored target data in the library. This eliminates the time needed for the human crew to identify the target, time in which the craft is vulnerable to enemy fire. The detection and classification system also helps identify and prioritize targets for attack. Other sensors include low-light-level TV, image intensifiers and a laser range-finder/designator. The advanced cockpit management system organizes and displays intelligence data and allows rapid digital transmission of critical battlefield information to the tactical field commander. Everything is connected via a fiber-optic sensor-data distribution network.

The Comanche is the Army's first all-composite helicopter using a composite box-beam spine that carries the load, rather than a load bearing skin. The frame carries the unloaded skin allowing for access holes and an ability to maintain airworthiness even within battle. The all-composite helicopter is field repairable. And it does all this without a weight penalty. It is crashworthy and ballistically tolerant for gunfire up to 23 millimeters. And if the Comanche does take a hit, separate redundant systems like the triple redundant electrical and hydraulic systems, and triple redundant digital fly-by-wire flight controls keep the RAH-66 flying. Also there are self-healing electronics, and the electronics are also hardened against disruptive electromagnetic interference.

Requiring fewer maintenance people was also an important consideration in the RAH-66's design. For example, all replaceable units are placed "one deep" so a good component doesn't have to be removed to get to the defective one. There are many doors and access panels, a feature made possible since the RAH-66 does not have a semi- or full-monocoque structure so a large number of cutouts for maintenance do not decrease structural strength. Only six tools are needed to remove any of the replaceable parts on the engine in under 15 minutes. Servicing can be done from ground level or standing on parts of the structures, like the open missile bay doors, without ladders or work stands. The only field tools needed are the 12 hand tools provided in a kit, and cotter pins. Safety locking wires or torque wrenches are not needed. All the diagnostic and prediction software and equipment are built in. The Portable Intelligent Maintenance Aid (PIMA), which looks like a laptop computer, is connected to the aircraft's avionics and stays with the Comanche both on the ground and in the air. The PIMA combines the features of a computerized technical handbook, electronic logbook that records all maintenance actions, and an analyzer that diagnoses inflight failures, displaying them once on the ground. The expert system function predicts battle damage effects on flyability and recommends appropriate remedies.

With external mounting, the RAH-66 can carry up to 14 AGM-114 Hellfire laser-guided missiles, 18 AIM-92 Stinger heat-seeking missiles, or sixty-two 2.75-inch rockets. The aircraft's 20 millimeter turreted Gatling gun has a firing rate of up to 1500 rounds per minute with a 500-round-basis ammunition load. Three soldiers can refuel and rearm a Comanche in fewer than 15 minutes even under the harshest conditions.

The Comanche RAH-66 is easily transported by any U.S. Air Force transport aircraft. A C-130 aircraft can carry one RAH-66, a C-141 can carry three, a C-17 can carry four, and eight can be loaded on a C-5. Disassembling a Comanche to load in a C-130 or C-141 takes 20 minutes and it takes 15 minutes to get ready for a trip in a C-17 or C-5. When equipped with external fuel tanks, the Comanche can be flown to trouble spots around the world within a 1260 nautical mile range.

V/STOL AIRCRAFT

For almost as long as man has wanted to fly, he has also wanted to take off and land just like a bird. This ability was finally achieved with the first successful helicopter flights in the 1940s. As useful as they have become, helicopters are limited in their maximum forward speed (about 200 knots) and their range. So there have been many attempts to develop other types of Vertical/Short Takeoff and Landing (V/STOL) aircraft. But the field has been littered with more failures than successes. Like so many other mechanical things that are called upon to do multiple tasks, most V/STOL designs were not able to do every task very well. They also turned out to be quite complex in terms of both hardware and the skill needed to fly them. Also because of complexity, the basic craft turned out to be quite heavy, leaving little capacity for useful payloads. Because of the military's potential use of V/STOLs, however, a rather intensive development effort has continued, and this investment in resources is starting to pay dividends.

The V/STOL fighter is especially attractive for close-support missions. Instead of having to loiter in the air expending fuel and being exposed to enemy fire while waiting to be called for fire support, it can wait on the ground at primitive air bases close to the battle zone. Because it can operate close to the battlefield, long flyback distances are not required, making high sortie rates possible. V/STOL fighters are able to operate from bomb-cratered runways. The Navy can fly a V/STOL from much smaller ships than giant aircraft carriers (Fig. 7-10). Because of its ability to fly slow and hover, the V/STOL fighter could provide a unique capability in

Fig. 7-10. With V/STOL aircraft, small ships like the one in this artist's concept could replace giant aircraft carriers.

General Dynamics.

urban-type warfare, such as destroying sniper nests located in the upper stories of buildings.

Aircraft that combine the best features of the fixed-wing airplane and the helicopter fall into two categories: aircraft designed primarily for high-performance flight that can also take off and land vertically as well as hover, and aircraft that are basically fast helicopters. The former is best represented by the AV-8 Harrier which has already proven itself in battle. The Bell Boeing V-22 Osprey tilt-rotor concept is the latest of the aircraft in the latter category.

One of the disadvantages of the AV-8B Harrier is that it is not capable of supersonic speeds. Fortunately, developing a supersonic V/STOL is technically feasible, the most significant problem being the development of engines with sufficient thrust to accelerate the aircraft beyond Mach 1 and still be able to provide vertical lift. Because of the success and versatility of vectored thrust as used in the Harrier, the major engine manufacturers such as Rolls-Royce, General Electric, and Pratt & Whitney are working on engines that could power a supersonic V/STOL. The interest in these developments has been piqued because of the supermaneuverability it could provide future fighters, in addition to V/STOL capability (Fig. 7-11).

The ability to take off vertically represents a large weight penalty in the V/STOL fighter. A modified version of the concept, the Short Takeoff and Vertical Landing (STOVL) aircraft would use a short rolling takeoff, perhaps assisted by

Fig. 7-11. Through the years, various proposals have been offered for a V/STOL fighter that could also fly supersonically. Here are some of NASA's concepts.

such devices as a "ski-jump" launching pad. After its mission is completed, with its fuel and ordnance expended, it would return for a vertical landing. Just how dramatic the increase in payload can be by using a short rolling takeoff is demonstrated by the Harrier. If a 1200-foot ground roll is used, as opposed to a vertical takeoff, the maximum gross weight of the Harrier increases by about 50 percent, and most of this extra capacity can be used for fuel and ordnance.

Tilt-rotor V/STOL

A military tilt-rotor V/STOL, the V-22 Osprey, has been developed by Bell Helicopters Textron and Boeing Vertol (Figs. 7-12 and 7-13). The V-22 used much of the experience gained with the Bell XV-15, which has been flying for well over a decade. The V-22 made its maiden flight in March of 1989. The Osprey is a hybrid aircraft; it can take off, land, and hover like a helicopter, and by tilting its rotor system forward, can also fly like a turboprop airplane at speeds up to 300 knots or 345 MPH.

In the helicopter mode, the V-22 is flown like a normal helicopter using collective and cyclic-pitch controls contained in the proprotor hubs. As a helicopter it can fly up to 200 knots (230 MPH) and sideward or rearward at 35 knots (40 MPH). Then during conversion from helicopter to airplane mode, the flight controls

Fig. 7-12. Among its many capabilities, the V-22 Osprey can be refueled in flight to extend its range so it can be self-deployed to trouble spots around the world.

Fig. 7-13. The V-22 can carry cargo from a sling or it can be used for rescues, just like a helicopter.

automatically transition to conventional elevators, rudders and flaperons to control pitch, yaw, and roll in fixed-wing flight.

For short-range combat missions, the V-22 can carry 20,000 pounds of cargo internally, and has an amphibious assault range of over 575 miles carrying 24 fully-equipped people. When carrying fewer troops or less cargo, the range increases to 1150 miles, about twice that of an ordinary helicopter. There is room for 12 litters for medevac missions. The Osprey is the worlds only airplane equipped with the external cargo hooks that are so familiar on helicopters; actually it has two cargo hooks plus a rescue hook. Using a dual-hook suspension system, the craft can carry up to 15,000 pounds suspended from the dual cargo hooks and at speeds of 230 MPH. The maximum range, with auxiliary fuel tanks, is about 2400 miles, allowing self deployment to trouble spots around the world. With a ceiling of 26,000 feet, it can fly much higher than most helicopters.

The Osprey is powered by twin Allison Gas Turbine Division 6150 shaft-horsepower, T406-AD-400 turboshaft engines connected to the proprotors. The two propulsion systems are cross-connected to allow emergency single-engine operation. If one engine's output falls below a preset limit, a torque sensor automatically engages the driveshaft that connects the two gearboxes, one for each engine. Immediately power is transferred from the other functioning engine so both proprotors can provide balanced thrust without interruption.

Several advances in technology are found aboard the Osprey. For example, the V-22's structure is made almost entirely of graphite/epoxy solid laminates. Only 1000 pounds of metal is used, mainly fasteners and copper mesh laminated into the outside surfaces for lightning protection. The three-bladed proprotors are made from advanced fiberglass and graphite materials. Besides resistance to corrosion, the composites have super-ballistic tolerance, are easy to repair in the field, and cracks do not grow as they do in metal structures. Also aboard are computerized digital avionics and a fly-by-wire flight control system. The avionics and flight controls are triple redundant. In the cockpit, full-color displays have replaced gauges and illuminated pushbuttons are used in place of toggle switches. All flight instruments, engine parameters, radio frequencies, and system status information are displayed on four multifunction full-color cathode-ray tubes (CRTs). The cockpit is pressurized for nuclear, biological, and chemical protection. There are provisions for night vision and adverse weather operations. Night and poor weather capability is also enhanced by a turreted Forward Looking Infrared Radar (FLIR) system.

While the U.S. Marine Corps is the primary customer for the Osprey, it was designed to meet the needs of all the military services. For example, the Marine Corps would use the craft for land and amphibious assault and support missions. Naval roles include combat search and rescue, special warfare, and fleet logistics support. The Air Force could use it for its long-range special operations, and the Army's tasks include medevac, combat support, and its special brand of operations.

X-wing

Another V/STOL concept is the X-wing, also referred to as the "stopped-rotor" because of the way the rotor, used for helicopter-like flight, is stopped in flight to serve as wings for fast forward flight. One major obstacle facing the designers was how to keep the aircraft from rolling over during the 30-second period when the rotor stops (or starts) during the transition between helicopter and fixed-wing modes. The answer was circulation control, often referred to as the Coanda principle. This consists of air forced through openings in the leading and trailing edges of the airfoil. The amount of air forced out determines the circulation around the wing and the lift produced. By finely controlling the circulation, stable flight can be achieved even during transition.

Circulation control also replaces much of the mechanical complexity found in helicopter rotor systems. With the X-wing as currently envisioned, it would be used for cyclic lift control, eliminating the mechanical pitching of the rotor blades during each cycle of rotation. In fixed-wing flight, it would be used for pitch control in conjunction with a horizontal stabilizer.

The entire propulsion system for the X-wing is a challenging design. Not only must it provide power to rotate the rotor, but also supply thrust for forward flight and compressed air for circulation control.

An X-Wing aircraft could be used by the Navy for anti-submarine, airborne electronic warfare, electronic intelligence, and combat search-and-rescue missions; missions that are typically conducted at sea, either from aircraft carriers or other surface ships. The V/STOL capabilities permit operations to be completely done from smaller ships, which frees up the carriers to support attack and fighter aircraft, and disperses the V/STOLs over a larger area of the ocean closer to their targets. Speed and range are important for these missions. The X-wings, with forward speeds in the high-subsonic region, aircraft-like ranges, and hover and vertical flight capabilities, could be just what is needed for future defense needs.

The Canard Rotor/Wing (CRW), proposed by McDonnell Helicopter, is one variation of the X-wing concept (Fig. 7-14). The CRW uses a stoppable rotor/wing plus both a canard and horizontal tail, all providing lift. While in the helicopter mode, both exhaust and bypass gases from the turbine engine are ducted through the blades and ejected through nozzles near the blade tips. These jets rotate the rotor/wing and eliminate the need for conventional transmission and anti-torque systems, a great savings in weight. The rotor/wing uses a two-bladed teetering rotor system with collective and cyclic pitch control. When converted to fixed-wing flight for high speeds, the turbofan engine provides forward thrust just like in a regular jet aircraft, with the stopped rotor acting like a regular wing. As the CRW increases speed, the canard and horizontal tail produce more lift, and less lift is required from the rotor. At 120 to 150 knots, there is no longer any need for rotor thrust and thus the rotor/wing can be locked in place. All three surfaces provide lift during fixed-wing flight. The rotor could be oriented obliquely for flight at high subsonic speeds. During fixed-wing flight, control of the aircraft is provided by active controls on the canard and horizontal tail which would both incorporate high-lift surfaces on their trailing edges. The CRW could be used on either a manned or remotely-piloted aircraft such as the Army's Future Attack Air Vehicle, the U.S. Marine Corps' Attack Observation Aircraft, or the multiservice VTOL Unmanned Aerial Vehicle.

Another interesting V/STOL concept is the M-85 being proposed by NASA's Ames Research Center (Figs. 7-15 and 7-16). The designation comes from the fact

Fig. 7-14. Artist's concept of an unmanned aircraft using a Canard Rotor/Wing (CRW). Here the rotor is being used as a wing for high speed forward flight.

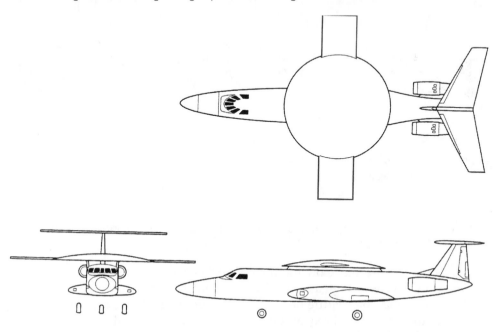

Fig. 7-15. Light M-85 transport shown with the hub fairing and two blades deployed as wings for flight at speeds of up to Mach 0.85. NASA.

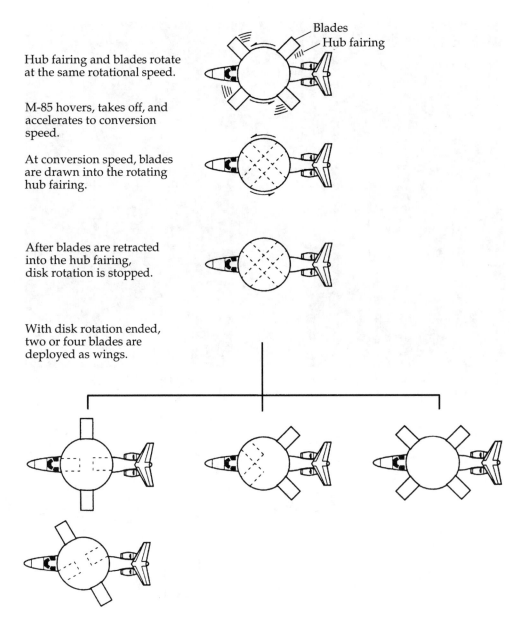

Hub fairing and blades rotate at the same rotational speed.

M-85 hovers, takes off, and accelerates to conversion speed.

At conversion speed, blades are drawn into the rotating hub fairing.

After blades are retracted into the hub fairing, disk rotation is stopped.

With disk rotation ended, two or four blades are deployed as wings.

Blades

Hub fairing

Fig. 7-16. Sequence of operation for the M-85 concept that can take off like a helicopter and eventually transition for flight at high subsonic speeds. NASA.

that while the M-85 can take off and hover like a helicopter, it can also reach subsonic cruising speeds as high as Mach 0.85 or about 450 knots at cruising altitude.

The most noticeable feature of the M-85 concept is the large, rotating circular hub fairing to which two, three, or four rotor blades are attached. The hub fairing is large enough, about 50 to 60 percent of the total rotor diameter, to produce sufficient lift to support the aircraft while it is converting to and from the rotary-wing and fixed-wing flight modes. This allows the rotor blades to be unloaded, that is, not producing lift, during conversion, greatly reducing vibration and oscillatory loads.

In one concept, during conversion the rotor blades could be retracted into the hub, the left blade would be rotated so its leading edge is pointed towards the direction of flight, the rotor is stopped, and then the blades would be extended out of the hub again. While the design could use two, three, or four blades for rotary flight, only two would probably be deployed for fixed-wing flight. Furthermore, the blades could be swept rearward, or even forward, for high subsonic speed. Alternately, the blades could be left extended during conversion. A control system would reorient the blades so leading edges were pointed in the correct direction with little or no lift being produced during the conversion.

Unlike a helicopter which is tilted downward so the rotor blades can provide forward thrust for acceleration, the M-85's circular hub is always maintained at a positive angle-of-attack, just like a fixed-wing aircraft. If the circular hub flew at a negative angle-of-attack it would produce negative lift. Therefore, thrust for acceleration comes from fan jet engines mounted along the fuselage, most likely located at the tail end.

Calculations show that during cruise speeds the large circular hub fairing, augmented by the rotor blades which are used like normal lifting surfaces, result in cruising efficiencies that are competitive with normal fixed-wing aircraft. During hover, the circular hub fairing reduces rotor thrust and hover efficiency only slightly compared to a pure helicopter. Maneuverability and handling approaches that of an airplane while in fixed-wing mode. Conversion between flight modes can be done in turbulence and even while maneuvering.

The rotor blades rotate at the same speed as the hub fairing and could be conventional or symmetric airfoils, or even use circulation control to produce additional lift. While operating as a helicopter, the blade pitch is separately controlled by rotary actuators eliminating the complexity of the normal helicopter-type swash plate. The hub could be rotated by a conventional shaft drive or by warm or hot pressurized air jets emitted from the blades or from the disk itself.

NASA says this concept could be used for a variety of different aircraft, from high-performance fighters to commuter-type commercial airliners. It would have a minimum speed of 160 knots in fixed-wing mode to allow a safe landing should it not be possible to convert to the rotary wing configuration.

AIRLIFT TRANSPORTS

The best military strategies and tactics are of little value if the right soldiers, weapons, and supplies cannot be in the right place at the right time. This is where strategic and tactical airlift aircraft come into play. Throughout the history of air power, airlift aircraft have not always received the attention they should have, perhaps because they are not as glamorous as fighters and bombers and do not usually push the state of aviation technology. Also, in many cases, the military has simply depended on slightly modified versions of civilian commercial airliners to do the job. This has been the case from the Douglas DC-3 which became the C-47, to the Douglas DC-10 which is currently on active duty with the Air Force as the KC-10. With the scarcity of resources projected for the future, this situation will probably not change.

Aircraft like the Lockheed C-5, C-141, and the C-130 will see many more years of service. The first C-130 flew in 1954, and new C-130s are still rolling off the production line. Different versions of the C-130 have ranged from gunships to stretched commercial versions. C-130s are popular aircraft with military and commercial operators around the world. Thus C-130s will still be a familiar sight in the twenty-first century, as will the giant C-5 which will still be needed to carry out-sized cargo, like Army tanks and howitzers, around the globe. Be assured, however, that while the basic appearance of these aircraft will remain constant, the interior, especially the avionics, will be continually updated (Fig. 7-17). For example, Litton Systems Canada Ltd. has modified the cockpit of a C-130 with Active Matrix Liquid Crystal Displays (AM/LCD) in an evaluation program conducted by the USAF. This "glass cockpit" replaced 60 electromechanical flight instruments with six 6 × 8-inch color displays.

Litton Systems Canada Ltd.

Fig. 7-17. The venerable C-130's cockpit updated with a "glass cockpit" that has been equipped with Active Matrix Liquid Crystal Displays (AM/LCD).

Even though these current airlifters will remain in the inventory for many years, the C-17 is vitally needed to airlift troops and cargo long distances and deposit them directly at the front. C-130s can land on small, austere fields but do not have the needed range. Although the C-141 has a longer range than the C-130, it requires longer, concrete runways. Neither of these can handle much of the Army's or Marine Corps' outsized equipment, but the C-5's huge size requires larger runways and ramps where aircraft can be parked while unloading. This means that C-5s and C-141s would carry cargo to larger fields in the combat zone, but equipment would have to be shuttled to the front by C-130s. The C-17 can bring outsized loads to three times more fields than can the C-141 or C-5, many of them located in Third World countries. The C-17 needs only three crewmen versus the five-to-seven needed by the C-5 or C-141 (Fig. 7-18).

McDonnell Douglas.

Fig. 7-18. The C-17 has the ability to carry tanks and other heavy military vehicles right to the battlefield.

While an advanced design, the C-17 does not really use any major technologies that have not been used before. The C-17 just uses them in a different way. For example, less than 10 percent of the C-17's airframe uses high strength composites, and then they are used in secondary structure control surfaces rather than primary load-bearing structures. The low usage of composites can be attributed partly to the fact the C-17 was designed in the mid-1980s. A quad-redundant, digital fly-by-wire flight control system includes a mechanical backup capability. The Delco mission computer, the brains of the C-17, links the avionics, controls, cockpit displays, and mission-related functions. The C-17 is the first military airlifter to feature a head-up display (HUD), one made by GEC Ltd. in Great Britain.

The C-17 will be able to drop supplies or to land on unprepared landing strips as short as 3000 feet. It will be able to do this after dropping rapidly from altitude, a maneuver used to reduce its vulnerability to ground fire. High-lift devices on the wings allows the C-17 to fly steeper approaches with no-flare touchdown velocities of up to 15 feet per second. Descents of 10 feet per second are considered hard landings for many aircraft, not just transports. The HUD allows a high 5-degree glideslope compared to about three degrees used by airline pilots. The C-17's four Pratt & Whitney F-117, 41,700 pound thrust turbofan engines and the wings are integrated into a high-lift system that permits them to fly steep, slow descents. The engines are mounted on wing pylons so that the exhaust exits relatively close to the underside of the wing. When the flaps are lowered from the wing's trailing edge, additional "powered lift" is generated by the exhaust blowing over the flaps. Each flap is a single movable surface. There is a large fixed slat mounted to the flap ahead of its leading edge, forming a slot. Therefore, the flap, once lowered, acts as a double flap to produce greater lift.

Exhaust from both the engine turbine core and the fan can be reversed, with the thrust directed forward and upward. This reduces the amount of dust and debris kicked up on unimproved fields. Thrust reversers allow the C-17 to maneuver, even back up, on tight ramps—even those with grades of up to 2 percent. Reversers can be deployed in flight to quickly descend from high altitudes.

For the future, there is also an interest in V/STOL airlifters that would require less than 300 feet for takeoffs and Super Short Takeoff and Landing (SSTOL) craft that require 1500 to 3000 feet, but when the situation dictates, could use 300 feet or less by activating auxiliary thrust boosters.

LIGHTER-THAN-AIR CRAFT

The military has been interested in lighter-than-air (LTA) craft—airships, balloons, aerostats, blimps, Zeppelins, and so forth—for decades, indeed even centuries. The Montgolfier brothers experimented with balloons in France more than 300 years ago, and there are many interesting tales of airships used during the Civil War, World War I, and World War II. The airship community prides itself with the fact that during World War II, no Allied convoy protected by an airship lost a single vessel to hostile action. The largest blimps ever built were the Navy's non-rigid, 403-foot-long ZPG-3Ws, which contained 1.5 million cubic feet of helium. They were an integral part of the North American Air Defense Command's warning system for detecting a possible Soviet bomber attack between 1957 to 1961. Using the then latest in piston engine technology and a large part of their payload (including their airborne-warning radars) carried within the helium compartment, their design has not yet been surpassed.

During the 1980s, the U.S. military services showed a renewed interest in airships, with Congress even appropriating money for LTA research. All the military services investigated LTAs for a variety of missions. LTAs were especially touted over other land-, air-, sea-, or space-based systems when the mission involved long-duration, low-altitude surveillance. The U.S. Navy was interested in advanced airships for detecting low-altitude cruise missiles because airships can overcome limitations of even powerful shipboard radars or normal aircraft for detecting low-flying missiles over the horizon. Huge airships, perhaps twice the size of the ZPG-3Ws, were considered to house the large, long-range, missile-detection radar needed for detecting supersonic and low-flying missiles which have minimum radar cross-sections). Also considered were "conformal" radars with their components being integral parts of the airship itself.

The Air Force's interest went far beyond studies and prototypes with its Tethered Aerostat Radar System (TARS) actually reaching operational status. While first intended to surveil our southeastern coast for possible military intrusion, the TARS system is now used by both the Air Force and the U.S. Customs Service for drug interdiction and surveillance. The system of unmanned airships now provides an "electronic picket fence" of virtually unbroken coverage to detect both airborne and maritime drug smugglers along the southern border from the Atlantic to Pacific oceans.

Incidently, an aerostat is an airship with two compartments separated by a bladder. The upper chamber contains helium and the lower one contains air. As the aerostat climbs, the pressure outside becomes less. Thus, the helium expands and forces air out of the lower chamber through electromechanical valves. As the aerostat descends, the helium compartment contracts and electric blowers pump air back into the lower section. This maintains both the aircraft's buoyancy and its aerodynamic shape at all altitudes. The TARS craft operates at an altitude of 10,000–12,000 feet, tethered by a cable that also transmits power and information from the on-board radar system. The tether is attached to a diesel-powered, locomotive-sized, launch-control vehicle that runs on a circular track. The aerostat is kept at its proper station by moving the control vehicle around the track and reeling the tether in or out.

The U.S. Coast Guard was also interested in smaller airships, perhaps in the neighborhood of 250,000 cubic feet, that could accommodate a small crew on two-day missions like rescues at sea or for transporting emergency gear to ships in distress. The Coast Guard was also interested in unmanned "aerostats" for some of its law enforcement activities. Finally, the U.S. Army has investigated the use of small, remote-controlled airships carrying a variety of sensors for security surveillance and damage assessment.

Today, Westinghouse Airships, Inc.'s Sentinel with its 353,100 cubic foot envelope volume is not only the largest airship flying, but it incorporates the latest in LTA technology (Fig. 7-19). For example, its twin, high-efficiency engines drive ducted propellers which can rotate independently through a total of 210 degrees to provide vectored thrust. Vectored thrust in an airship allows, for example, near vertical takeoffs, greatly reducing the space required for a ground run, plus enhanced low-speed stability during ascent, descent, and hover even under light wind conditions.

The Sentinel 1000 is the first "aircraft" to use a fly-by-light control system with fiberoptics transmitting commands from pilot to control surface. Among other advantages, fiberoptics reduce problems caused by electromagnetic fields. While

Fig. 7-19. While looking like airships of days past, the Sentinel 1000 uses much advanced technology including a fly-by-light control system and advanced materials like Tedlar for the envelop.

there are two seats in the cockpit, the Sentinel is flown by a single pilot using a side-stick controller. Avionics include digital instrumentation and a three-axis autopilot with full instrument flight capability ensuring low pilot workloads for long-endurance flights.

The outer layer of the helium filled envelope, an ideal enclosure for sensors and antennas thus reducing drag, is made of a high-tech material called Tedlar that is impervious to damaging ultraviolet light and many other environmental elements. A hangar is needed only for airframe maintenance over the expected 30 year lifetime. Besides being very strong and light, the envelope is radar transparent making it naturally stealthy. The gondola holding mission equipment, cockpit, crew quarters, and the galley is made of Kevlar, and the tail fins use glass fiber composites. The 220-foot long Sentinel 1000 has a top speed of 67 MPH, can stay aloft for up to 30 hours, and has a maximum altitude of 10,000 feet.

The Sentinel 1000 is really a half-scale prototype of the Sentinel 5000, a craft being studied by the Department of Defense as part of its Air Defense Initiative. However, smaller LTAs like the 1000 could be used for a variety of future military missions. Modern airships are exceptionally stable, low-altitude platforms that could carry or tow sensors for mine countermeasure work at sea. It has the relatively long endurance, augmented by the ability to refuel at sea, as well as low-speed controllability, allowing it to hover over slow moving convoys. Because economic sanctions and blockades are becoming a routine foreign policy measure, the smaller airship could be used in naval blockade work to continuously detect and track hostile country commerce from point of origin.

UNMANNED AIR VEHICLES

Throughout history, soldiers have wanted to know what was happening behind enemy lines before engaging the foe in battle. Scouts and spies have ventured behind enemy lines to gather intelligence, but the military has also known for centuries that you can get the maximum amount of information and a feel for the "big picture" if you do your spying from a high place, preferably the sky. Thus, balloons were used during the Civil War, and the airplane's first mission was aerial reconnaissance during World War I. Aerial reconnaissance was quite routine during World War II and reached a zenith with the U-2 and SR-71 aircraft.

Manned reconnaissance aircraft are expensive, however, not only in terms of risking lives, but also in their design and testing. Thus, unmanned air vehicles (UAV) have advantages, especially aircraft that can be controlled in flight, namely remotely-piloted vehicles or RPVs. For many years, the Air Force has had a rather successful RPV that even saw action in Vietnam. This RPV, of which there were about two dozen different versions, was the Teledyne Ryan AGM-34. Its missions included low-level day and night reconnaissance, high-altitude surveillance, and electronic intelligence. It and the SR-71 were the only air vehicles allowed to fly over North Vietnam for reconnaissance after the cessation of bombing in January of 1973. The AGM-34s were either ground-launched or launched from an aircraft.

The Israelis have also been a leader in using UAVS for military missions. For example, RPVs were used in the 1973 Arab-Israeli War. More recently, RPVs, made to look "electronically" like Israeli aircraft, were used to keep Syrian gunners busy while Israeli reconnaissance airplanes flying at higher altitudes collected intelligence data on Syrian missile sites. On other occasions, while Syrian radars were tracking the RPVs, the Israelis fired anti-radiation missiles at ground-missile sites by homing in on the radar signature. They also used RPVs for near-real-time reconnaissance of vital ground targets destined for air attacks. The Israelis found RPVs to be essentially immune to enemy action because of their small size and minimum radar signature.

American UAVs proved themselves during the Persian Gulf War. Five short-range (about three miles) Pointer and six longer-range (over 100 miles) Pioneer units were deployed during the conflict. Of the six Pioneer units, two were deployed on Navy battleships, three with Marine Corps companies and one with an Army platoon. Almost 50 Pioneers flew more than 530 sorties for nearly 1700 combat hours. At least one Pioneer was airborne at all times during the war and only one was lost to enemy fire. Over land and sea, the Pioneers performed a variety of missions including surveillance and reconnaissance, mine detection and disposal, spotting for artillery and naval guns, advanced warning, operations coordination, and real-time damage assessment.

One of their primary missions was to penetrate enemy territory, replacing human scouts, forward observers, and patrols. UAVs were able to fly under clouds, dust, and smoke to provide views not obtainable from higher flying manned spy aircraft and satellites in space. One Pioneer was probably the first robot in military history to capture humans when a group of Iraqi soldiers surrendered as a Pioneer flew by and the crew of the USS Missouri, the Pioneer's home ship, watched on TV monitors. Of perhaps equal importance was the experience gained from using UAV under real battlefield conditions, information that will be invaluable in designing and developing the unmanned aircraft of the future (Figs. 7-20 and 7-21).

Fig. 7-20. You can see why the CL-227 remotely piloted helicopter has earned the nickname "Peanut."

Fig. 7-21. The Pioneer Unmanned Aerial Vehicle (UAV) is used to provide real-time intelligence and reconnaissance for the field commander. It flies at up to 115 MPH and can stay aloft for over five hours.

Another UAV, the Navy's air launched ADM-141 Tactical Air Launched Decoys (TALD) were used to draw enemy anti-aircraft fire. These UAVs, about half the size of a manned tactical aircraft, flew into Iraqi airspace at Mach 0.8 emitting both passive and active radar signatures. Likewise the Air Force launched its unmanned BQM-74 Chukar that could fly all the way to Baghad on its jet engines. The Iraqis turned on their anti-aircraft radar in response to the decoys and the radar was almost instantanously destroyed by HARM anti-radar missiles launched from Navy F/A-18 and USAF F-4G Wild Weasel aircraft.

Military planners have come to realize that UAVs can supplement or enhance, but will not replace manned aircraft. Both manned and unmanned platforms are needed to give commanders flexibility for reconnaissance and surveillance missions. Even pilots, usually not proponents of systems that could put them out of a job, recognize and support their value for high risk missions against integrated air defenses, on missions lasting more than 24-hours, and for missions where there is potential exposure to chemical or biological agents. RPVs can sneak up on the enemy because of their small size and essentially noiseless operation. Low radar and infrared signatures make them hard to detect, even by sophisticated sensors. By using RPVs, pilots do not fly over high-threat environments or incur the political implications of having a manned craft shot down over a hostile or neutral nation. Downed pilotless aircraft leave neither widows nor POWs (Fig. 7-22).

Advances in miniaturized electronics and sensors are important to UAVs because of their usually limited payload capacities. State-of-the-art devices like very small day-and-night TV cameras, FLIRs, radar, laser designators, and navigation and communication systems will allow UAVs to be small enough to be carried and

Fig. 7-22. The TRW/IAI Hunter is another UAV that can be used for real-time surveillance, communications, and control of the battlefield.

flown by even a single soldier. Future RPVs will be easier to control from ground using autopilots, inertial navigation systems, and the Global Positioning System (GPS) to keep the RPV at the right altitude and attitude, and on the right track to and from the target. The future RPV "pilot" would only have to command the RPV when unprogrammed changes are required.

The disadvantage of fixed-wing RPVs like the Pioneer is their need for a complex launcher and a runway, net, or parachute for recovery and reuse. Incidently, while Pioneers were designed to land on runways, during the Gulf War they were routinely landed on ordinary roads. One solution to the problem of a lack of runways in battle situations is the remotely-piloted helicopter or RPH. One example of an RPH is the Canadair CL-227 which can take off and land vertically, as well as fly at horizontal speeds of up to 80 MPH. This RPH uses two contrarotating rotors, eliminating the need for a tail rotor. Its hour-glass shape keeps radar detectability to a minimum, as do its radar-absorbing body materials and Kevlar rotors. The CL-227 can be launched and recovered from a truck or even the deck of a small ship.

Another approach is to use an unmanned V/STOL aircraft using tilt rotor technology. This is the idea behind the Bell Helicopter Eagle Eye Unmanned Aerial Vehicle (Figs. 7-23 and 7-24). The Eagle Eye would operate like the V-22, but would be unmanned and remotely controlled. Such an UAV could be used for a variety of military missions like over-the-horizon surveillance, battle damage assessment, and electronic countermeasures. The craft would have a composite airframe.

High costs have been one of the biggest impediments to RPVs being used to their fullest potential. Advocates often tout the RPV as a "low-cost" solution to

Bell Helicopter Textron.

Fig. 7-23. The Eagle Eye Unmanned Aerial Vehicle uses a tilt rotor to operate both as a helicopter and a fixed-wing aircraft.

military problems. But experience has shown otherwise. When designs of RPVs begin, they usually have low-acquisition and operational costs as goals. However, during development things happen to drive costs up. First of all, the military services want RPVs to do a multitude of jobs. Then they demand super reliability and want to use people with a minimum of experience and training to operate them. All this adds up to complex, heavy, sophisticated, and expensive RPVs. As they become very valuable, they must be made reusable, and this increases the complexity and cost of the ground equipment. Before you know it, system costs of the UAV approach those of the manned system it was designed to replace. At that point the military often would rather have a manned system.

Now, with low-cost miniaturized electronics, low-cost RPVs can be designed

Fig. 7-24. The Eagle Eye Unmanned Aerial Vehicle under construction. Note the all-composite structure and the single turboshaft engine that drives both proprotors.

and built. But the military manager must make sure that the project is kept on course and be sure not to add "nice but not necessary" capabilities, or to try to accomplish a multitude of different missions. Perhaps an RPV could be designed for only a few uses or could even be expendable. In the heat of battle, a commander would probably rather have many simple RPVs at his disposal versus only a few expensive ones. The loss of one or two of the latter would represent a significant portion of his assets. Also because RPVs are unmanned, less stringent requirements should be placed on componentry. Simplicity reduces electrical power requirements which have a direct bearing on system weight.

The cost of air vehicle and ground equipment could, however, be better amortized over a large number of multipurpose units. Different black boxes or cassettes could be inserted into the same air vehicle, depending on the mission. (Note that this is different from a single, complex RPV that could do many things.) There would be different payloads for day and night reconnaissance, as well as a different module for laser target designation. Payload module changes would be simple enough that they could be done right in the field.

8

Commercial aircraft

THE AIRLINE INDUSTRY'S ACCEPTANCE of new aircraft and new technology is keyed to profitability and safety. Profitability can come from both reduced operating costs and extra services for which a substantial number of people are willing to pay a surcharge. Safety can be actual or perceived. Airline travelers may not want to fly in airliners that, despite their safety, look too unusual. Air cargo, on the other hand, is not as choosy. So air-freight and air-express carriers might be the most influential force in commercial aircraft design of the twenty-first century. Air cargo could well be the reason for building aircraft with unique designs optimized for carrying cargo, as well as loading and unloading freight.

CONVENTIONAL AIRLINERS

The McDonnell Douglas MD-11, Boeing 777, and Airbus Industrie A330 and A340 are the latest conventional air transports. These are the aircraft that will carry a huge number of passengers and volume of cargo well into the twenty-first century.

The McDonnell Douglas MD-11, a wide-cabin, triengined aircraft has been in airline service since 1991 (Fig. 8-1). Like most commercial air transports it is built in several configurations, an all-passenger airliner for 250 to 410 passengers, a freighter with over 100 tons of payload capacity, and a convertible freighter that can serve either as an airliner or an air freighter after a quick change or "combi" where passengers and freight are carried on the top deck, and more freight is carried on the lower deck. Three General Electric or Pratt & Whitney turbofan engines can propel the MD-11 to a maximum speed of 588 MPH, or Mach 0.87, and to distances of almost 8000 miles.

Aerodynamic features such as winglets, redesigned wing trailing edges, smaller horizontal tail with integral fuel tanks, and an extended tail cone add up to reduced drag saving fuel and increasing range. On the flight deck, the crew views operational, navigation, engine, and systems on six color cathode ray tubes (CRT) (Fig. 8-2). Other advanced systems include wind-shear detection capability, a dual flight-management system that helps conserve fuel, and a dual, digital automatic flight-control system or autopilot. Computerized system controllers perform automated, normal, abnormal, and emergency checklist duties for major systems, reducing flight crew requirements from three to two.

Fig. 8-1. McDonnell Douglas MD-11 Trijet comes in three configurations—all-passenger, all-freighter, and "combi" (or convertible) freighter.

Fig. 8-2. The flight deck of the McDonnell Douglas MD-11 features six CRTs.

The Boeing 777 is a huge twinjet with seating capacity ranging from 250 to 440 passengers (Fig. 8-3). Like other Boeing airliners, "stretched" versions could appear in the future. A longer-range (7600 versus 4660–4900 miles), higher-weight version is in the plans as well. The initial Boeing 777 can be fitted with turbofan engines built by General Electric, Pratt & Whitney, or Rolls Royce.

Boeing Commercial Airplane Co.

Fig. 8-3. The Boeing 777 is a huge twin-engined airliner.

The 777 uses many advanced materials like improved 7055 aluminum alloy in the upper wing skin, and stringers and carbon fiber composites in the vertical and horizontal tails. Floor beams in the passenger cabin are made of advanced composite materials, as are secondary structures like aerodynamic fairings. In all, 10 percent of the 777's structural weight is composites.

The 777's cockpit contains state-of-the-art, flat liquid-crystal panel displays (Fig. 8-4). Six large screens present principle flight, navigation, and engine information. The flat displays save space since they are half as deep as a comparable CRT, and they weigh less and consume less electrical power. Because they produce much less heat, they are more reliable, have longer lives, and do not require heavy-duty air-conditioning systems. They are easily read under all conditions, even in direct sunlight.

The twin-engined Airbus Industrie A330 and four-engined A340 look very much alike. That is because they share widebody fuselages, wings, major components, and advanced on-board control systems. Indeed, while having different

Fig. 8-4. The flight deck of the Boeing 777, featuring liquid crystal displays. .

missions, about 90 percent of the two aircrafts are similar. The A330-300 is a relatively short-range (less than 5000 miles), high-capacity airliner while the A340-200 is designed for longer routes (about 7500 miles) with a somewhat lesser number of passengers (Fig. 8-5). The common features of the airliners not only reduced development and production costs, but also led to significant savings in maintenance costs because common procedures, spare part inventories, training, and so forth can be used. Also, because flight decks are virtually identical (except for four throttle levers on the A340 versus only two on the A330), flight crews can easily transition from one aircraft to the other.

The Airbus aircrafts use fighter-like side-stick controllers that provide an unimpeded view of the six CRT displays plus other instruments. An on-board centralized maintenance system connected to all of the aircraft's principle systems records any malfunctions and provides a single source for troubleshooting and testing.

Like all Airbus airliners since the A320 airliner (in service since 1988), the A330/A340 uses a digital fly-by-wire control system. Airbus Industrie, a consortium composed of Aerospatiale, Deutsche Airbus, CASA, and British Aerospace has much experience with fly-by-wire systems. The Concorde used an analog fly-by-wire system and the earlier A300-600 and A310 had limited fly-by-wire control systems. The A330/A340 uses five main computers, any of which can fly the aircraft, plus mechanical controls for backup as well.

Fig. 8-5. The Airbus Industrie Ultra-High Capacity Aircraft (UHCA) could carry up to 600 passengers to distances of up to 7000 nautical miles/13,000 kilometers.

The A330 and A340 use several techniques to minimize fuel consumption including drag-reducing winglets and center-of-gravity management that were pioneered on the Concorde. These techniques pump fuel between tanks in the wing and the tail to achieve optimum trim. Significant weight savings come from extensive use of composite materials. For example, the mainly carbon fiber vertical fin is not only 20-percent lighter than one made of aluminum alloy, but has only 100 parts, versus about 2000 in a comparable metal fin, thus reducing production costs. The A330/A340's wing/fuselage belly fairing is made of a hybrid of carbon fiber and fiberglass, and is the largest composite structure (in surface area) used on any current airliner. Other composite components include the horizontal tailplane, also used as a fuel tank, access panels, and landing gear doors. In all, 15 percent of the airframe weight is composite materials.

The A340 is quite capable of flying 14 hours or more without landing. Thus, two crews may be required. The Airbus can be fitted with rest compartments as an option (Fig. 8-6). Here there are two bunks, wardrobes, and a complete in-flight entertainment and communication system for the off-duty crew. Another option is a "business center" with telephone, fax, and telex facilities enabling business travelers to keep in full contact with the world while airborne.

Like other air transport builders, Airbus Industrie has additional versions in mind. For example, longer distance A340s would carry over 260 passengers

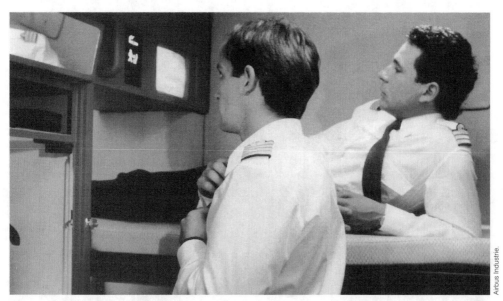

Fig. 8-6. The crew of the A340 could use a dedicated rest compartment on low duration flights when two crews are needed.

between destinations of nearly 9000 miles allowing nonstop flights from, for instance, Paris to Singapore, or even Paris to Sydney. Stretched versions are also in the works with the capacity for a twin-engine A330 carrying nearly 380 passengers, and a four-engine A340 carrying 335 passengers. Increases of about 40 people for both aircraft are achieved by adding about 21 feet to the common fuselage.

JUMBO JETLINERS

Some studies have shown that there will be a sufficient market for jumbo jets capable of carrying 600, 700, or even 800 passengers by the beginning of the twenty-first century. Much of this new market will be in the fast-growing and profitable Pacific Rim and Asian routes. The results of studies have shown 50 pairs of cities that could economically justify a 600-seat aircraft.

Therefore, several aerospace companies are investigating new aircraft to meet these projected needs. For example, McDonnell Douglas has developed its MD-12 (Figs. 8-7 and 8-8). This four-engined aircraft could carry more than 600 passengers in its double deck within a fuselage three feet wider than the Boeing 747. Boeing is looking at ways to make the big 747 jumbo jet even bigger by increasing its current 412 passenger capacity to as many as 550 passengers. Airbus Industries with its Airbus Ultra High Capacity Aircraft (UHCA) study, is also looking into an airline that can carry 600 passengers. Here passengers could be carried within several possible fuselage configurations, such as a double-decker with circular or egg-shaped cross-sections, or even a horizontal, "double-bubble" with two cylinders joined side-by-side. The craft could be as large as one-and-a-half times the size of today's largest airliner, the 747. Future jumbo jetliners could feature sleeper berths,

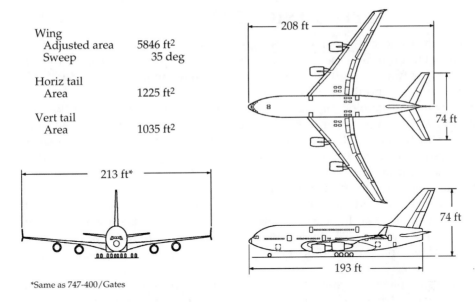

Wing
   Adjusted area   5846 ft²
   Sweep   35 deg

Horiz tail
   Area   1225 ft²

Vert tail
   Area   1035 ft²

208 ft

74 ft

213 ft*

74 ft

193 ft

*Same as 747-400/Gates

Fig. 8-7. The McDonnell-Douglas MD-12 Jumbo Jet. McDonnell Douglas.

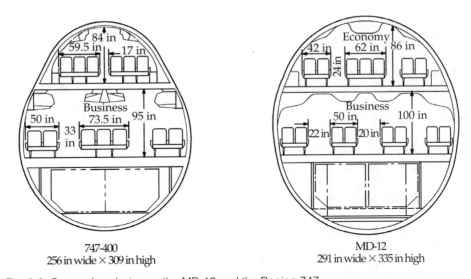

747-400
256 in wide × 309 in high

84 in
59.5 in — 17 in
Business
73.5 in — 95 in
50 in
33 in

MD-12
291 in wide × 335 in high

Economy
42 in — 62 in — 86 in
24 in
Business
50 in — 100 in
22 in 20 in

Fig. 8-8. Comparison between the MD-12 and the Boeing 747. McDonnell Douglas.

a restaurant, cocktail bars, business centers, observation lounges, exercise rooms, and even showering facilities.

Widespread use of jumbo jets would require some significant infrastructure changes. Most important is modifying or building airports that not only can handle the huge aircraft, but also the hundreds of people, baggage, and cargo each one carries. Just the ability to evacuate hundreds of people in an emergency from a disabled aircraft in less than 90 seconds represents a challenge.

SUPERSONIC TRANSPORTS

Except for the Concorde, previous SSTs were stopped from flying by economic and environmental concerns, noise pollution, and sonic boom problems. These problems still thwart any future SSTs. Thus the Concorde is limited in the routes it can fly, the airports it can use, and the speeds at which it can fly over land. This British-French venture, while a technological success, has been pretty much a financial flop. The Concorde's Rolls-Royce Olympus turbojets (essentially 1960s fighter engines), while fairly efficient at cruising altitude, use huge amounts of fuel and are extremely noisy during takeoff and climb. This resulted in relatively poor range and even required additional fuel tanks to be installed, which reduced passenger capacity. Both of these factors cut deeply into the Concorde's ability to produce revenues.

A future SST would have to overcome some pretty significant hurdles, both fiscal and technological. To be a business success, some studies show that a future SST would have to carry 250 to 300 passengers at Mach 2 to Mach 3, cover twice the range of the Concorde, and do it for one-seventh the cost per passenger mile—a pretty tall order.

One of the biggest obstacles to a successful future SST, and the factor that would probably drive the design of its propulsion system, is the environmental concern about the nitrogen oxides, or NO_x, exhaust emissions. Current theory indicates that excessive NO_x emissions could severely damage the earth's atmosphere by contributing to the deletion of the earth's ozone layer. Depleting the ozone layer allows more harmful ultraviolet radiation to reach the earth's surface. Some experts say that a fleet of SSTs using conventional jet engines might reduce global ozone by as much as 15 percent. Future SSTs would fly in the stratosphere, well above altitudes used by current airliners, where NO_x is particularly damaging to the atmosphere. Fear of ozone layer depletion has already brought a worldwide effort to ban ozone-layer depleting chemicals like chlorinated fluorocarbons (CFCs).

In order to achieve the higher efficiencies needed, SST engines would operate at much higher temperatures. Every 200 degrees (Fahrenheit) increase in inlet air temperature doubles the amount of NO_x produced. The higher combustion pressures also needed add even more NO_x emissions. Fortunately scientists and engineers believe the problem is solvable and are already working on advanced engine designs. These engine designs optimize such items as the fuel/air mixture, engine temperatures, and airflow velocities through the engine to reduce NO_x emissions while increasing engine efficiency. But an ultra-low-NO_x engine is still a long way off. Another approach being considered is to use NO_x reducing additives, probably in combination with other approaches. The goal is quite ambitious since emissions probably have to be reduced five to tenfold to reduce the ozone layer loss to only a few percent. Finally, the SST could fly at lower altitudes since ozone depletion ceases to be a problem below about 40,000 feet. However, the much denser air would significantly increase drag, meaning only lower supersonic speeds could be used, which decreases the SST's advantage over subsonic airliners.

As for the problem of sonic booms, researchers are looking at various approaches, including shifting the bulk of the acoustic energy to the lower frequency range since these frequencies are less annoying to humans. However, most of the problem will be solved, if it can be solved at all, by choosing the proper shape of the SST. For example, sonic boom could be reduced by a long needle-shaped fuselage with lifting surfaces starting near the nose and gradually increasing in area while blending into the main wing. This shape would distribute aircraft volume and lift over the widest area to minimize boom energy produced.

Engine noise during taxi, takeoff, and climb is another challenge. Possible solutions include internal engine design changes that reduce turbine noise by mixing quieter, low-velocity outside air with the high-velocity hot gases exiting the turbine. Another approach could be a variable-cycle engine that combines the relatively quiet turbofan for lower-speed, lower-altitude operation with a turbojet mode for efficient high-altitude, supersonic cruise. Some of the noise can be reduced by careful design of the engine inlets and nozzles. Also, by using high-thrust-to-weight characteristics in conjunction with automatic throttle controls, the SST could accelerate faster to its best climb speed, and thus exit the airport area sooner. Likewise, descents during landing approaches could be steeper. While noise on the runway would still be higher, noise offensive to the airport's neighbors might be less than today's airliners. Engineers are investigating more sophisticated flight techniques that could result in substantial noise reduction. Unfortunately, many noise reduction approaches can result in performance penalties in terms of thrust lost or increases in fuel consumption. These trade-offs make the search for quieter engines even more difficult.

A future supersonic transport could travel at speeds of Mach 2.7, making a trip from Los Angeles to Tokyo in a little over three hours. To reduce fuel consumption, the factor that plagues the Concorde's economy, a very slender body design could be coupled with a low-aspect-ratio wing to greatly reduce wave drag. For further drag reduction, active laminar-control techniques, such as those that suck off the boundary layer, could be used. Wing-body blending, as found on military aircraft, would also improve aerodynamics (Fig. 8-9).

Advanced high-temperature-resistant, low-weight material are key ingredients for a successful and profitable SST. Composites can decrease the weight of an aircraft by as much as 20 percent compared to conventional metal-alloy construction. An SST would use extensive amounts of composites, such as high-strength carbon fibers embedded in a polymide and bismaleimides polymer matrix, as well as high-temperature aluminum alloys. As speeds increase, more exotic materials are needed in areas subjected to high levels of aerodynamic heating. Thus, the materials list for a SST will include items like metal-matrix composites made of silicon-carbide fibers embedded in rapidly solidified aluminum and titanium-matrix composites.

Another idea for an SST is the oblique-wing concept explained in Chapter 3. According to proponents, the oblique wing could provide optimum performance under all SST flight conditions.

Through the years, several oblique-wing airliner concepts have been proposed. The latest oblique flying wing idea eliminates the conventional fuselage by having passengers ride inside the wing. The seats, arranged in short rows, would face sideways, perpendicular to the flight direction. There would be windows in

Fig. 8-9. Artist's concept of the Alliance, a proposed supersonic transport to replace the Concorde.

the leading edge of the wing (Fig. 8-10). With a wingspan of about 500 feet, it could carry up to 500 passengers. Computer-controlled flaps on the thin trailing edge provide pitch and roll motions. Yaw steering control is provided by a pair of vertical rudders located near the wing tips. The oblique-wing airliner could be propelled by a pair of large, high-thrust turbofans that use a variable-bypass-ratio design. Using engine technology employed in the Advanced Tactical Fighter (F-22), the engines would operate very efficiently over a wide range of speeds and altitudes. Without the need for afterburner, the aircraft would be relatively quiet even when accelerating to its maximum speed of around Mach 1.6. The aircraft could change the amount of obliqueness by swiveling the engines on their pylons, and the rudders would swivel as well. Automatic stability control not only allows this basically unstable aircraft to fly well, but allows the wing to be loaded from front to rear with fuel, passengers, avionics, and other mechanical equipment.

Computations show that an oblique-wing airliner could fly at Mach 1.6, twice the speed of the Boeing 747, while consuming no more fuel than a subsonic jumbo jet. This level of efficiency is achieved because of the oblique wing's very high lift-to-drag ratio and, subsequently, relatively low thrust requirements, even during takeoff and acceleration. Initial wind tunnel testing has shown the oblique wing could have a takeoff lift-to-drag ratio as high as 30:1, that is 30 pounds of lift for every pound of aircraft weight. This is because there is basically no load on the structure most of the time since the gravity load is offset at every point on the aircraft by the lift. By comparison, conventional airplane wings act as long levers requiring massive internal structures to handle huge bending loads caused by having to support a heavy central fuselage.

Fig. 8-10. The oblique all-wing airliner could fly at speeds of up to Mach 1.5. At this speed, the wing would be angled at 70 degrees from the position of a normal straight wing (which would be used for landing and low speed flight).

At a sweep angle of 45 degrees, the oblique flying wing will be able to fly as fast as 750 MPH without any sonic boom. Beyond this, sonic boom elimination is not really possible, thus higher supersonic flight would have to be done over water or by overflying the Arctic region.

LARGE COMMERCIAL TRANSPORTS

While very large passenger airliners capable of carrying many hundreds or even thousands of passengers might be hard to justify and fund, there is definitely a market for very large cargo aircraft. These air freighters would fill the gap between today's cargo aircraft and surface ships.

While aircraft will never be able to replace or even economically compete with huge-capacity surface ships, they do offer the advantage of speed. Even the slowest of the proposed huge aircraft provides a tenfold increase in speed over surface craft, an important consideration where cargo is perishable, of high value, or must be delivered to market rapidly. Most large-aircraft concepts are being proposed for international and transoceanic transportation.

Through the years many unusual designs have been proposed to increase the efficiency and capability of commercial transports, especially cargo carriers. None went beyond the very initial concept stage. Perhaps they will be pulled out of the drawer again in the future, and maybe even built!

The spanloaders are a variation of the flying wing, chosen now because of its cargo carrying efficiency (Fig. 8-11). As the gross weight of an aircraft grows,

Fig. 8-11. A spanloader air freighter would allow rapid loading and unloading.

longer and thicker wings are needed, and bending loads at the wing roots (where the wings and fuselage meet) become tremendous. The wings become thick enough to provide cavernous cargo compartments and eliminate the need for a fuselage. By distributing the load over the entire wingspan, bending loads are substantially reduced (thus, the term "spanloader"). Eliminating the fuselage also reduces drag since fuselages normally are drag producers and contribute very little to the overall lift of an aircraft. While in today's air freighter only about 10 to 25 percent of the total gross weight is revenue-producing cargo, in a very large spanloader cargo could account for as much as 50 percent of the total takeoff weight. The greatly improved structural efficiency would come not only from the distributed wing loading, but also from the extensive use of composite materials. Active control systems would be used to reduce stresses during turbulence and maneuvering. Supercritical wings would also probably be used, their thick characteristics adding to the payload capacity. Typically, spanloaders would be propelled by six, eight, or even more engines. To speed up cargo handling, they could be loaded through doors located in the wingtips. Because of their huge wingspan—350 to 500 feet—they could not operate from conventional runways, which are neither wide enough nor sufficiently strong to bear the huge load. One idea is to use special hub airports to handle spanloaders. Smaller feeder aircraft would deliver the cargo to and from the hub. Other ideas include air-cushion landing gear in place of wheels. Riding on a cushion of air that distributes the load over a large area, it could land and taxi on soft ground, ice, or snow.

One way of getting around the runway problem of a large aircraft is to use multibody aircraft with two or three fuselages joined by main and tail wings (Fig. 8-12). The multiple-fuselage aircraft could increase the structural efficiency of spanloaders by distributing weight better than a conventional single-fuselage aircraft. Only the outer wings are cantilevered while the wing sections between the fuselages are simply supported. The cargo would be carried in the fuselages rather than the wings, which would be reserved mainly for fuel. The multibody air freighters would not be as aerodynamically efficient as the spanloaders because they would still have drag-producing fuselages, but they would be less expensive to develop and build. Even fuselages from currently produced airliners could possibly be used. Loading of two or three fuselages might be accomplished in a shorter time than just one fuselage holding the same cargo volume. Incidently, catamaran-type aircraft are not a new idea. The twin-hulled Savoia-Marchetti SM-55 and SM-66 were built in Italy in the 1930s, and after World War II, two P-51 fuselages were joined together to make the very-long-range F-82 fighter.

NASA.

Fig. 8-12. A dual-body cargo carrier could use twin fuselages originally designed for normal airliners.

The flatbed transport concept consists of an abbreviated fuselage that essentially contains only the crew compartment. Behind this is the cargo area that, in its basic form, consists of a completely flat area. Outsized cargo such as heavy machinery or construction equipment would simply be loaded on the flatbed, lashed down, and flown exposed (Fig. 8-13). Naturally, the drag would be higher than with a streamlined fuselage, but that is the penalty to be paid for carrying cargo that cannot be loaded within a conventional aircraft fuselage. Containerized cargo would be loaded directly on the flatbed without an outer covering. To reduce drag, when the containers are fitted together they would function as a streamlined fuselage, eliminating the need for a conventional fuselage and the resultant weight penalty.

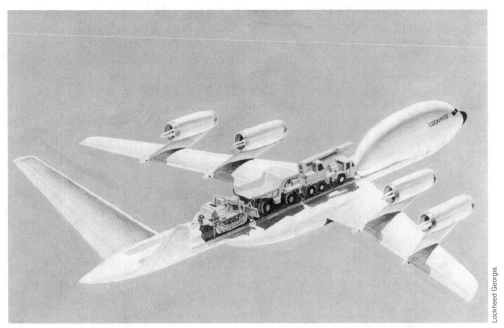

Fig. 8-13. Outsized cargo could be carried out in the open on a flatbed air freighter.

For cargo requiring a pressurized or temperature-controlled environment, a removable "cocoon" could be fit over the containers, or the containers themselves could be insulated and pressurized. The flatbed transport could even be converted to a passenger-carrying airliner by adding a passenger module on top of the flatbed which could be removed and wheeled to a passenger terminal for loading and unloading. Such an idea would reduce terminal congestion and mean smaller airport terminals because the aircraft could stay out on the flight line. The passenger module could even be loaded on wheels and driven on the road or over a track to provide transportation between the airport and the city. The flatbed would be a truly efficient convertible airliner, something that airline operators have sought for years. It could haul passengers during the peak daytime hours and freight at night. And the conversion could take place very rapidly.

An aircraft flying very close to the surface of the earth gets a sizable boost in lift from the cushion of air compressed between the wings and surface. However, because this surface effect extends only to an altitude of about one-half the wingspan, the wingspan would have to be large enough to permit the aircraft to fly at a safe distance above the surface. Thus, the wing-in-ground effect (WIG) is best suited to very large aircraft. And because the only location where long-range travel can be made over smooth surfaces is over oceans, a WIG aircraft would also have to be a seaplane.

While the ground effect of a large WIG aircraft would be felt a hundred feet or so "off the deck," the most dramatic increase in lift occurs if the aircraft flies at an altitude of 10 to 20 feet. This might seem a bit dangerous, but the surface has a stabilizing effect. If the aircraft flies too low, the lift increases, causing the aircraft to climb a bit. If the altitude becomes too great, lift decreases and the aircraft settles down. Likewise, in a roll the low wing has increased lift and the high wing has less lift, so the aircraft automatically returns to a level position. To reduce the loss of lift due to tip vortices, end plates which operate like winglets would be used. These end plates would literally skim over the tops of the waves.

One wrinkle to the WIG idea is the addition of a power-augmented-ram (PAR) propulsion system. This consists of several large turbofan engines mounted on the forward part of the fuselage. During takeoff, these engines would be tilted so that the exhaust would raise the pressure under the wings, increasing lift and decreasing takeoff distance. While the PAR-WIG aircraft would operate most efficiently at extremely low altitudes, the aircraft could even fly over land at normal altitudes. For flight at low altitudes, speeds would be kept relatively low; for example, speeds would be less than 400 MPH.

PAR-WIG aircraft with gross weights of over two million pounds and payload capacities approaching one million pounds have been considered in preliminary design studies (Fig. 8-14).

Lockheed Georgia.

Fig. 8-14. The wing-in-ground effect (WIG) transport would probably operate at an altitude of around 100 feet from the ocean's surface.

9

General aviation aircraft

IT IS EASIER to say what general aviation is not, than to define what it is. General aviation covers every type of piloted airborne craft that is not in service with either the military or the major commercial airlines. Even commuter, air-taxi, and private business aircraft are included. Not only does general aviation have the largest number of aircraft, over 220,000 in the United States at last count, it has the greatest diversity in technology as well. General aviation aircraft range from light aircraft that were built with pre-World War II technology, to future supersonic business jets that have been proposed by a couple of aerospace companies.

Staggering lawsuit settlements and liability insurance premiums have probably hit the general aviation market harder than any other segment of society, except, perhaps for the medical profession. By 1986, the manufacturer's product-liability insurance cost per-new-aircraft was averaging $70,000, up from a mere $2000 in the early 1970s. This has not only brought general aviation aircraft production to almost a standstill, but has greatly curtailed applicable research and development. Legislative action, namely tort reform, and not technology will be the solution to the problem.

As you will see, general aviation enthusiasts and many manufacturers have found ways to get around these problems, and these solutions will set the tone for general aviation well into the twenty-first century. One method is to update older, less liability-prone designs with the latest aerodynamics, avionics, engines and so forth.

The ultimate of these updated aircrafts is probably the Basler Turbo-67 (Fig. 9-1). Basler Turbo Conversions, Inc. of Oshkosh, Wisconsin is making "all-new" DC-3s out of old ones. Of the 10,000 plus DC-3s and military C-47s built before production ceased in 1946, it is estimated that between 500 and 1000 still exist, either still flying or in storage.

The new, updated DC-3s are retitled the Basler Turbo-67, the name coming from the twin Pratt & Whitney Canada PT6A-67R turboprop engines that replace the normal DC-3's piston engines. Performance increases include a maximum cruise speed of 205 knots versus the DC-3's 170 knots. The conversion also includes a 40-inch fuselage stretch and the cockpit bulkhead is moved ahead by some 60 inches to provide 35 percent more cargo carrying capacity. The useful load increases from about 9000 pounds to nearly 13,000 pounds. The DC-3's wings are

Fig. 9-1. The Basler Turbo-67 "looks like an old one, but flies like new." By adding turboprops and many other modifications, the venerable Douglas DC-3 and C-47 gets a new lease on life.

beefed up for the additional load and modifications are made to wing tips and leading edges to improve flight characteristics.

Every wire, hydraulic line, hinge, and fitting is replaced. The flight deck is completely redone with the latest avionics and instruments. This aircraft comes with a "zero time since overhaul" rating and a new aircraft warranty (Fig. 9-2).

Fig. 9-2. DC-3s and C-47 being converted to Basler Turbo-67 at the company's facility in Oshkosh, Wisconsin. With up to 1000 of these nearly half century old airplanes still existing, there is lots of "raw" material for the conversion.

BUSINESS AIRCRAFT

Executives and celebrities demand air transportation that meets their personal needs; they do not want to tailor their schedules around the timetables of the commercial airlines, or they want to fly to the small surburban or rural airstrip closest to their destination. This has created a flourishing demand for executive transports. And if hijackings and terrorism continue, important people will find that only private air transportation can provide the security they desire.

Because of this expanding market, new business aircrafts are starting to roll off the production lines, with many more on the drawing board and in various stages of development. Many will also serve in the commuter airline and air-taxi markets, but with more spartan interiors than the plush executive transports.

When it comes to commercially-built business aircraft, few are as advanced as the Beechcraft Starship 1 (Fig. 9-3). With its canards, tipsails, and all-composite construction, it looks like an aircraft for the twenty-first century, but is already flying as the world's first pressurized all-composite business turboprop. Twin, 1200-shaft horsepower Pratt & Whitney PT6A-67A engines, turning the five-bladed McCauley propellers, are mounted in the rear to reduce cabin noise as well as provide safer control if one engine should quit. Because a conventional rudder would have made a huge sounding board for the propellers, the Starship uses large tipsails on each wingtip. Tipsails not only provide directional stability, but also reduce drag by reducing the wingtip vortices.

Beech Aircraft Corporation.

Fig. 9-3. The Beechcraft Starship looks like an aircraft we would expect to see flying in the twenty-first century, but it is already flying.

The Starship 1's tandem-wing design uses two lifting surfaces, the front canard and rear main wing. Unlike conventional aircraft where the wing lifts and the tail provides a balancing downforce, the tandem-wing configuration provides lift for maximum combined lift with minimum drag. It also gives the Starship much more docile stall characteristics and improved ride qualities over conventional designs. The Starship's canard (Beechcraft calls it a variable forward wing) has computer-controlled variable geometry so that it can sweep farther aft for less drag at high-speed cruise or sweep forward to provide greater stability at low speeds. It also allows extension or retraction of flaps without having to make trim changes. The bottom line is that the Starship 1 uses at least one-third less fuel than business jets (Fig. 9-4).

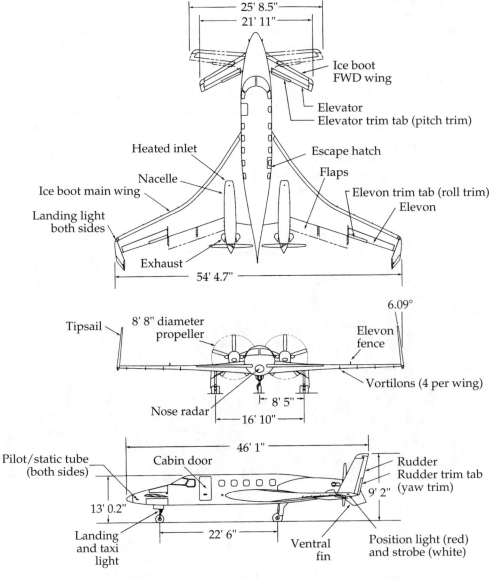

Fig. 9-4. Tipsails, pusher turboprops, all composite airframe, and moveable canard are just some of the Starship's advanced technology. Beech Aircraft Corporation.

The Starship 1 has a maximum cruise speed of 335 knots (385 MPH) and is certified for operation with two engines at altitudes of up to 41,000 feet. It requires only 2630 feet to land and just over 4000 feet to take off with a maximum weight of 14,500 pounds, 4500 pounds of which is useful load. Depending on the altitude and gross weight, the Starship's range can be as great as about 1600 miles.

The high-tech Starship 1 is constructed of a non-metallic, Nomex honeycomb core sandwiched between thin facing sheets of graphite/epoxy. Beside being approximately 15-percent lighter than if the aircraft had been built of conventional aluminum construction, there is much greater resistance to fatigue and corrosion. Because of the composite's high strength-to-weight ratio, not only is the Starship lighter, but the cabin walls are only 2.5-inches thick, about half the thickness of a comparable aluminum aircraft. This permits a more slender fuselage with less drag and without sacrificing interior room. Indeed, the Starship 1's cabin width approaches that of medium-sized business jets costing twice as much.

Protection from lightning strikes is a real challenge in designing and certifying an all-composite aircraft. Compared to aluminum, graphite epoxy is about 1000 times more resistant to the flow of electric current. Current flowing through high-resistance materials results in excessive heat and temperatures, and, in the case of a lightning strike, this happens extremely rapidly. Unprotected composite material can actually be blown apart by a lightning strike. To provide the required lightning protection, the Starship uses a combination of fine wires in the first layer of composite skin and a ground-plane system to shield the electronics. This allows the lightning current to flow through and out of the aircraft, leaving only minor surface and cosmetic damage at the strike point.

The Starship 1's cockpit is as modern as the external design, and was designed specifically by Rockwell Collins Avionics. Stepping into the pilot's "office," you immediately notice the 14 CRTs used for the electronic flight information system (EFIS), sometimes referred to as a "glass cockpit." Most of the CRTs are full color with superb computer-generated graphics. The fully-integrated system provides significantly-reduced pilot workload, matching or exceeding the efficiency found in many EFIS-equipped airline cockpits, and exceeding the capability of comparable general aviation aircraft. Gone is the clutter found in most cabins. In its place is an efficient, workload-reducing presentation of information that allows the crew to make rapid decisions and keep a clear view of the outside world. Incidentally, without a large wing or engine nacelles to block the view, visibility out of the cockpit is excellent.

If you think the crew is treated well, you should see the accommodations for the passengers riding in the fully-pressurized cabin (Fig. 9-5). All the accoutrements of an executive suite are here—comfortable folding seats, premium stereo, flight phones, lavatory, and provisions for preparing food and refreshments. The cabin provides seating for eight passengers in comfort.

Business aircraft buyers have several other high-speed aircraft to choose from that incorporate much advanced technology. The Learjet 45 can carry eight in luxury at speeds of up to Mach 0.81 to distances of up to 2500 miles without a stop (Fig. 9-6). The turboprop Gulfstream V (see Chapter 1), with its 6300 nautical mile range and Mach 0.9 top speed, is the first of global business jets that will be fairly common in the twenty-first century.

Fig. 9-5. The Beechcraft Starship I offers a wide cabin despite relatively narrow external dimensions. Thin composite walls make this possible.

Fig. 9-6. Another business jet with the look of the future, the Learjet.

Overall, you might say when it comes to business aircraft, the twenty-first century has already arrived. However, with businesses taking on global dimensions, subsonic private "airliners" may not be fast enough for high-level executives who want to hold meetings halfway around the world and return home the same day. To meet their needs, manufacturers like British Aerospace, one of the builders of the Concorde, have explored the possibilities of a supersonic business jet that could cruise at around Mach 2 and have a range of 3800 miles. British Aerospace called its design concept the "Concordette" because it looks somewhat like its older brother (Fig. 9-7). There could be a substantial market for supersonic executive transports in the twenty-first century.

Fig. 9-7. The ultimate business jet, a supersonic concept from British Aerospace.

LIGHTPLANES

Of all aviation, the area that might be least affected by advances in aerodynamics, materials, and propulsion technology could be single-engine light aircraft built by the traditional aircraft companies—Piper, Cessna, and Beechcraft. Because of product liability problems, sales of new lightplanes will probably remain in their current doldrums unless changes in laws are made. Except perhaps for fancier paint jobs, the lightplanes commonly seen around airfields today look pretty much the same as the designs seen in the 1950s, as probably will be the case in 2001. While the traditional lightplane builders will probably stick to proven designs, technology, and construction practices, this will not be the case for avionics. Advances in electronics plus demands for greater safety will mean that, while lightplanes might represent 1950s' concepts, the instrument panel will keep pace with the progress of other aviation sectors.

Fortunately, innovations in the lightplane industry are not completely dead. Many smaller, entrepreneurial aircraft builders are continually developing some pretty radical lightplane designs, often incorporating concepts pioneered by homebuilders (Figs. 9-8 and 9-9). Unfortunately, economics often prevent most innovative designs from ever reaching production.

Fig. 9-8. The Seawind, a rather futuristic light seaplane from Canada.

The restoration and refurbishment of old lightplanes could allow the average aviation enthusiast to keep on flying. The cost of completely restoring an old airframe and engine to "like new" condition is substantially less than the purchase price of a brand-new airplane. In the case of the old airframe, product liability insurance and certification costs are almost nil. And if new avionics are added, plus perhaps an upgraded engine, the restored lightplane is just about as good as new because lightplane performance has not changed dramatically. This means that the demand for used aircraft will be tremendous, and aircraft restoration businesses could be very profitable. Investments in used aircraft and refurbishments could be wise investments because lightplanes will be appreciating in value. As a result, many aircraft that might have been relegated to the scrapyard may be given a new lease on life.

HOMEBUILTS AND ULTRALIGHTS

While rapid incorporation of technological advances might not be the way of life in the lightplane industry, this is definitely not the case when it comes to the air-

Fig. 9-9. A joined-wing recreational aircraft concept.

craft being assembled in garages, carports, and basements across America (Fig. 9-10). Indeed, new ideas and completely new designs are the mainstay of these experimental-aircraft builders. Homebuilding is a thriving hobby, as witnessed by the tens of thousands of people that flock to the Experimental Aircraft Association's annual air show in Oshkosh, Wisconsin.

Fig. 9-10. The Cirrus VK-30 shows the level of sophistication now available in homebuilt aircraft. The 4/5 seat VK-30 can cruise at speeds of up to 300 MPH when propelled by a turbocharged engine.

While many homebuilts look like future military and commercial aircraft, others resemble historic aircraft. These are the homebuilt, subscale versions of aircraft like the Spad or Fokker biplanes of World War I fame, or the P-51 Mustang from World War II.

Burt Rutan has had a major impact on homebuilt aircraft. His designs, like the VariEze and Long-EZ, show that canards can be successfully used on an aircraft the average pilot can safely fly. He pioneered the use of fiberglass-on-foam construction. This type of construction is simple and low-cost, but it is also strong and its smooth surfaces result in reduced drag, the idea behind the natural laminar flow concept.

Even though he did not invent the idea of subscale flying prototypes, Burt Rutan made them a useful tool in aircraft development. Prototypes like the AD-1 oblique-wing demonstrator (the subscale version of the Beechcraft Starship I) and the T-46A trainer allowed flight testing of new configurations at a fraction of the cost of a full-size flying prototype. By using composites, not only were the subscale prototypes built quickly and inexpensively, changes in designs were easily incorporated and tested.

V/STOL AIRCRAFT

While the primary impetus for the development of V/STOL aircraft will come from the military which is most able to fund their development, concepts like the V-22 Osprey tilt rotor could find many civil applications. Such an aircraft would be ideal in remote parts of the world where distances to be traveled are great and airfields are primitive or non-existent. One example that immediately comes to mind is an island-hopping service in the Pacific. In addition, they would be a boon to the offshore petroleum industry, one of the major civilian users of helicopters. As near-shore petroleum deposits are depleted, exploration and production will have to be done at ranges beyond the capabilities of helicopters. Because V/STOLs are faster, less time would have to be spent in commuting to and from the rigs.

The tilt-rotor air-taxi or feeder airliner could operate from rather small landing areas and still offer the speed and range of a fixed-wing, turboprop aircraft. A cargo version would be perfect for air-express carriers to pick up cargo in small cities and rapidly transport it to large hub airports for transfer to the larger, long-distance air freighters. A tilt-rotor air ambulance could save precious minutes transporting critically injured persons from the site of an emergency directly to the hospital.

UNMANNED AIRCRAFT

While unmanned aircraft have many military applications, pilotless aircraft could also be used for many civilian tasks, such as monitoring electrical power transmission lines and oil pipe-lines or patrolling natural borders and secure areas. Remotely-piloted aircraft could also perform dangerous civilian flying jobs such as crop dusting. In Japan, Yamaha has developed its R-50 unmanned, remote-control helicopter with belly mounted tanks that hold pesticides or fertilizer. With a maximum take-off weight of 148 pounds, the R-50 powered by a 98 cubic centimeter, 12-horsepower engine can stay aloft for about half an hour. The R-50 is used in Japan for crop dusting on small farms, flying at altitudes of about 10 feet.

Unmanned aircraft are especially attractive for scientific research missions. Scientists are investigating the earth's upper atmosphere to understand such phenomena as ozone layer depletion and global warming. Much of the scientists' attention is focused on what is happening in the rarified region from 10 to 20 miles (60,000 to 120,000 feet) above the earth's surface. The observations often have to be made over periods lasting days or weeks.

There are several ways to get scientific data from these regions but all have serious limitations. While balloons, the mainstay of atmospheric research for decades, can stay aloft almost indefinitely, they have rather random flight patterns because they are at the mercy of high altitude winds. Also balloons fitted with expensive scientific instruments are sometimes lost. Satellites are great for obtaining the "big picture," overall views, but cannot gather the detailed data needed because they travel far above, and not in, the region of interest. The high-flying, manned ER-2, a modified version of the U-2 spy plane, has already flown on many scientific missions. Its single-jet engine is limited to altitudes below about 13 miles (70,000 feet) and can stay airborne for a maximum of eight hours. Flights of the ER-2 in Arctic and Antarctic regions can be very dangerous if there are mechanical problems. The SR-71 Blackbird military reconnaissance aircraft can fly at higher altitudes than the ER-2, but at Mach 3 it heats the surrounding air and creates shock waves that destroy or scatter the very molecules the scientists want to study.

The answer to these problems is unmanned, remotely-controlled aircraft that could stay up for very long periods of days, or even weeks, and can travel thousands of miles without refueling. An example of such an aircraft is the Perseus-A, developed by Aurora Flight Sciences Corporation, the Perseus-A was based partly on experience gained from the Daedalus 88 human-powered aircraft that set several world records on a flight between the Greek islands of Crete and Santorini in 1988.

The Perseus is characterized by its pencil-thin fuselage and long, tapered wings, both made of Kevlar and graphic composite fiber materials (Fig. 9-11). The huge, 15-foot diameter propeller is needed to push the craft through the thin upper atmosphere. The 1300 pound Perseus is propelled by a 65-horsepower rotary engine that recycles its own exhaust, which is reused along with gasoline and oxygen stored aboard the aircraft. This technique, borrowed from powerplants designed for submarines and torpedoes, eliminates the need for heavy and expensive superchargers typically used for high-altitude flight. The Perseus, being proposed for NASA's Small High-Altitude Science Aircraft (SHASA) program can fly up to 82,000 miles with a 110-pound payload.

Let's look at a typical mission using the Perseus to map the suspected ozone hole. After launch, a technician flies the Perseus while it is still within line-of-sight communication. Then it spirals upwards using an onboard computerized navigation system tied in with the Global Positioning Satellites. Sensors in the front payload section seek out the tiny ice crystals of a polar stratospheric cloud, an indicator of ozone layer destruction. The aircraft flies a zig-zag course as the sensors determine the boundaries of the cloud and reverse the direction of flight to map the ozone hole. Then, out of fuel, the Perseus glides silently back to its home base.

Fig. 9-11. The Perseus is an unmanned scientific research aircraft that could be used to study the earth's upper atmosphere. Aurora Flight Sciences.

There are other aircraft on the drawing board that could accomplish high-altitude scientific missions. One concept, proposed by Endosat, Inc., would use electric motors supplied with electrical energy beamed via microwaves from portable antennas on the ground. In Germany, the Deutsche Forshungsanstalt fur Luft und Rahrfahrt (DLR) has proposed a two-man Strato-2C aircraft powered by twin-turbocharged engines that could fly up to 85,000 feet or 10,000 miles without refueling. There is also the Boeing Condor developed for the Department of Defense for secret missions (Fig. 9-12). This 20,000 pound craft, much larger and much more expensive than other proposed crafts, uses twin, liquid-cooled 175-horsepower engines driving 16-foot long propellers, and has already been demonstrated on flights up to 67,000 feet, durations of up to 2½ days, and distances of up to 20,000 miles.

LIGHTER-THAN-AIR CRAFT

Books and movies have been written about the era of the luxurious commercial airship which ended with the fiery crash of the Hindenburg. While passenger-carrying airships, except perhaps for tours and cruises, will probably not return, lighter-than-air craft (LTA) could be used for a variety of civilian tasks. Indeed, they are already being used extensively for aerial advertising and television coverage of sporting events.

As for the military missions discussed in Chapter 7, LTAs are logical candidates for jobs that require a long-time-on station and where quiet, unobstructive surveillance is important. For example, besides military roles, Westinghouse is proposing its Sentinel series of airships for fire-watch patrol and communications during natural disasters. Infrared sensors and low light level TV (LLTV) could be used to detect fires and direct firefighters to otherwise invisible hot spots. They can also provide communication relay in rugged and hilly terrain. The same platform and sensors could be used for other tasks, like traffic and disaster prepared-

Fig. 9-12. The Boeing Condor is an all-composite unmanned aircraft designed for autonomous, high altitude, long-endurance missions such as weather monitoring and atmospheric research.

ness, thereby providing the multimission capability that is desirable to amortize costs. Because LTAs are not as intrusive as helicopters, they are ideal for law enforcement work. Equipped with high-technology sensors like light-weight radar, forward-looking infrared (FLIR) camera, and other electro-optical systems, an airship could detect aircraft and boats at distances out to as much as 200 nautical miles in drug surveillance and border patrol duty. A LTA like the Sentinel could also be used for scientific missions such as observing the oceanic environment or for fishery protection (Fig. 9-13).

Fig. 9-13. The American Blimp Corporation's Lightship incorporates much advanced technology.

The American Blimp Corporation's Lightship represents the current state-of-the-art in low-cost commercial airships. It is designed to perform missions from law enforcement and disaster surveillance to aerial photography and mapping. The Lightship's design is aimed at simplicity in flight, ground handling, and maintenance. Its twin German Limbach, 68-horsepower L2000 EC1 engines turn 5-foot propellers burning only 4 gallons of fuel per hour while cruising. It can stay aloft for up to 15 hours, allowing it to operate from remote, unprepared sites. The 128-foot long Lightship Model A-60 has a 68,000-cubic-foot envelope and the gondola can hold the pilot plus four-passengers. The cockpit, while simple, allows day/night, VFR/IFR flight.

Luftschiffbau Zeppelin GmbH of Friedrichshafen, Germany, renowned for its pre-World War II zeppelins, has never completely abandoned the revival of a modern day airship. The company is currently developing its LZ N05 demonstration airship. Besides demonstrating new technology, the 206-foot long N05, could be marketed for scientific research, environmental monitoring, fishery surveillance, TV coverage, advertising, and traffic control. Zeppelin is also planning a much larger airship, the 360-foot long LZ N30, with an envelope volume of 30,000-cubic meters (1,060,000-cubic feet) versus only 5400-cubic meters (19,000-cubic feet) for the LZ N05. The latter could carry 84 tourists at speeds of up to 140 kilometers per hour (87 MPH), lift up to 15,000 kilograms (33,075 pounds) of cargo, or stay on station for as long as 82 hours performing long duration surveillance missions. Luftschiffbau Zeppelin has already built and tested a 10-meter long (32.8 feet) radio-controlled "proof of concept" model of the Zeppelin NT (Fig. 9-14).

Fig. 9-14. Remotely controlled scale model of the Zeppelin NT being used for testing new airship concepts. Luftschiffbau Zeppelin GmbH.

The NT, the letters standing for "New Technology," includes several new ideas and technology. For example, the envelope design includes features of both rigid airframe LTAs, like the zeppelins of old, and the differential-pressure techniques used in a multicell blimp to enhance the stiffening. The NT uses a primary internal structure extending over the entire length that consists of aluminum tubular longerons and carbon-fiber composite tubular frames arranged into a triangular structure. Aluminum was chosen for longerons because of its good elastic properties and ease in which it can be formed into an envelope contour that results in efficient aerodynamics at high cruising speeds. Carbon fiber was used for the frames because of its high specific strength and stiffness. Besides reducing the structural weight, the differential-pressure technique using multiple gas cells provides cushioning in a hard landing, and the rigid structure allows safe flight even if the differential pressure is lost (Fig. 9-15).

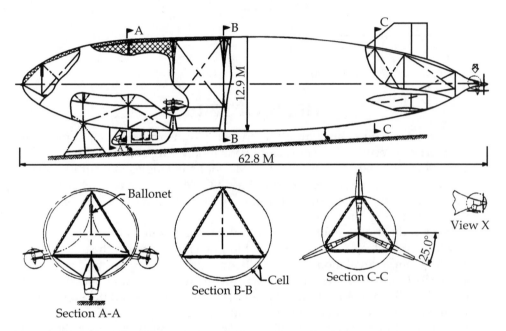

Fig. 9-15. The Zeppelin NT is a rigid airship like the Zeppelin of years past, but uses new materials and construction techniques. Luftschiffbau Zeppelin GmbH.

The NT design also incorporates vectored thrust engines, one on either side, and another at the stern. Vectored thrust would overcome the difficulties of ballast management and achieve more payload without sacrificing maneuverability. Thrust vectoring permits safer takeoffs and landings with a minimum ground crew. Incidentally, studies have shown that the majority of airship losses occur during takeoff or landing when wind gusts are common and the airship has little or no directional control other than ground crews pulling on lines. Vectored-thrust forces applied not only at the center of gravity but at the stern provides the control force needed at very low speeds. Speaking of control, the airship would use either a fly-by-wire or fly-by-light control system (Fig. 9-16).

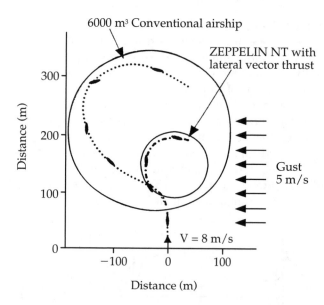

6000 m³ Conventional airship

ZEPPELIN NT with lateral vector thrust

Gust 5 m/s

V = 8 m/s

Distance (m)

Distance (m)

Fig. 9-16. By using thrust vectoring with a stern-mounted engine, the Zeppelin has significantly greater maneuverability compared to a conventional airship. This enhanced maneuverability is especially important during landing. Luftschiffbau Zeppelin GmbH.

HELICOPTERS

Helicopters are doing many important civilian jobs, from police and emergency service work to executive transport and traffic monitoring and reporting.

After some 17 years in development, McDonnell Douglas has put its NOTAR concept into production, the letters stand for NO TAil Rotor. NOTAR is available on the MD 520N five-place, single-turbine engine helicopter and the eight-place, twin-turbine MD Explorer (Fig. 9-17).

Instead of the traditional tail rotor to counter the torque of the main rotor and provide yaw control, NOTAR uses a variable-pitch fan located in the tailboom just behind the cabin, a circulation control tailboom, direct-jet thruster, and twin vertical stabilizers (Fig. 9-18). The fan, driven by the main transmission, pressurizes the air inside the tailboom. A portion of the low-pressure air is forced through the two slots running down the right side of the tailboom causing the main rotor downwash to "hug" the contour of the tailboom. This airflow pattern causes the downwash airstream to move faster around the right side of the tailboom compared to the left side. Like the airflow around a wing, the velocity difference results in a lift force, now sidewards, that provides the majority of the antitorque force required while hovering and at low speeds. If there were insufficient antitorque forces, the fuselage would rotate in a direction opposite to the main rotor's motion.

Additional antitorque forces and directional control are provided by the direct-jet thruster located in the tail. In an ordinary helicopter, the pilot varies the pitch of the tail-rotor blades to change their thrust, for example, to turn or compensate for crosswinds. In the NOTAR, this is done by the direct-jet thruster. This thruster consists of two concentric sets of air ducts. The inner one, which does not rotate, turns the air sideways. The outer one that does rotate, changes both the direction and the amount of air that leaves the thruster. By rotating the outer ducts

Fig. 9-17. The McDonnell Douglas Explorer with its NOTAR (NO TAil Rotor) antitorque system. The circulation slots can be seen along the right side of the tail boom.

using conventional rudder pedals in the cockpit, the pilot can obtain the precise amount of side thrust needed to fly the helicopter. At higher speeds, antitorque forces are countered by the tail-mounted vertical stabilizers which also provide yaw control.

There are several advantages gained by eliminating the tail rotor. First, the airflow from the tail rotor interacting with the main rotor flow can decrease the lift produced by the main rotor by as much as 20 percent. Therefore, if the tail rotor is missing, this loss in lift efficiency is also missing.

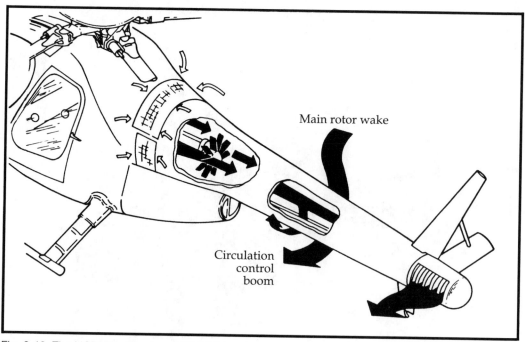

Main rotor wake

Circulation
control
boom

Fig. 9-18. The NOTAR helicopters (MD 520N and Explorer) use circulation control to eliminate the tail rotor. Circulation controls supply the necessary side (yaw) forces under most flight conditions.
McDonnell Douglas Helicopters.

Major improvements in safety are probably of greatest interest to many helicopter operators. There is no tail rotor to strike objects or to walk into, causing an accident. Without the danger of a tail rotor, the NOTAR helicopter can take off and land in much more confined areas, such as on a city street or at the scene of a highway accident without endangering ground personnel or civilians. They can also land in rugged locales covered with trees and brush with a reduced chance of being damaged or getting into a dangerous flight situation. Compared to normal helicopters that are limited to landing on slopes of no greater than about 10 degrees, the NOTAR helicopter can operate from slopes of up to 20 degrees, an important advantage in hilly or mountainous terrain.

Tail rotors, because they rotate at much higher speeds than the main rotor, are one of the main sources of high noise level. Without a tail rotor, noise is reduced by as much as 50 percent when compared to other competitive helicopters, making the NOTAR helicopter the quietest helicopter in the world. Quiet operation means reduced noise pollution allowing the helicopter to be a "good neighbor" to the residents of the city. In police work, the NOTAR could sneak up on suspects before they knew it was there because the helicopter noise would blend in with the noise of the street.

The NOTAR helicopter's high cruising speed, 140 plus MPH for the MD 520N and 170 MPH for the MD Explorer, results in faster response times. The MD 520N has a range of 380 to 400 miles, can stay airborne for about 2.3 hours, and can carry nearly a ton inside, or a bit more from an external hook. The larger MD Explorer has a 345 to 370 mile range, about 3.5 hours of endurance, and lifts almost 2600 pounds inside or 3000 pounds suspended from a cargo hook. The huge front canopies of both helicopters provide greater visibility for rescues or locating accident scenes.

Unlike helicopters with tail rotors, the NOTAR-equipped choppers require no more effort to hover upwind than downwind. Enhanced maneuverability comes with the ability to change headings at higher yaw rates and to bank at angles that would be difficult or even impossible for conventional helicopters. Also there is reduced vibration in the NOTAR because of the five-bladed rotor, which reduces vibration over two- or four-bladed rotor systems. The bottom line is reduced pilot fatigue and increased passenger comfort.

Epilogue

IT IS OUR FERVENT HOPE THAT THE ADVANCES IN technology described here will be used to benefit mankind. While many of the advances are aimed at increasing military might, civilian and military leaders must be resolute in using this technology to ensure peace and the preservation of freedom. This requires the delicate balance symbolized in the Great Seal of the United States, where the eagle clutches the olive branch of peace in one claw and a brace of arrows in the other claw, representing the military ability to ensure peace. This will not be an easy task, nor has it ever been. Christ told his disciples,

"And you will be hearing of wars and rumors of wars; see that you are not frightened, for these things must take place, but that is not yet the end." (Matt. 24:6)

Orville Wright best summed up our hope for aviation when, during the Great War, he said it was their hope that they were giving the world "an invention which would make further wars practically impossible." Unfortunately, history proved him wrong. The Wright brothers discovered that flight, like peace, requires not only brute power, but also control and balance. With this bit of wisdom from the first men to fly, may America and other nations of the world put into operation the best aerospace vehicles in the twenty-first century.

Index